Advanced Nuclear Radiation Detectors

Materials, processing, properties and applications

IOP Series in Emerging Technologies in Optics and Photonics

Series Editor

R Barry Johnson, a Senior Research Professor at Alabama A&M University, has been involved for over 50 years in lens design, optical systems design, electro-optical systems engineering, and photonics. He has been a faculty member at three academic institutions engaged in optics education and research, employed by a number of companies, and provided consulting services.

Dr Johnson is an IOP Fellow, SPIE Fellow and Life Member, OSA Fellow, and was the 1987 President of SPIE. He serves on the editorial board of *Infrared Physics & Technology* and *Advances in Optical Technologies.* Dr Johnson has been awarded many patents, has published numerous papers and several books and book chapters, and was awarded the 2012 OSA/SPIE Joseph W Goodman Book Writing Award for Lens Design Fundamentals (second edition). He is a perennial co-chair of the annual SPIE Current Developments in Lens Design and Optical Engineering Conference.

Foreword

Until the 1960s the field of optics was primarily concentrated in the classical areas of photography, cameras, binoculars, telescopes, spectrometers, colorimeters, radiometers, etc. In the late 1960s optics began to blossom with the advent of new types of infrared detectors, liquid crystal displays (LCDs), light emitting diodes (LEDs), charge coupled devices (CCDs), lasers, holography, fiber optics, new optical materials, advances in optical and mechanical fabrication, new optical design programs, and many more technologies. With the development of the LED, LCD, CCD, and other electro-optical devices, the term 'photonics' came into vogue in the 1980s to describe the science of using light in the development of new technologies and the performance of a myriad of applications. Today optics and photonics are truly pervasive throughout society and new technologies are continuing to emerge. The objective of this series is to provide students, researchers, and those who enjoy self-education with a wide-ranging collection of books that each focus on a relevant topic in the technologies and applications of optics and photonics. These books will provide knowledge to prepare the reader to be better able to participate in these exciting areas now and in the future. The title of this series is *Emerging Technologies in Optics and Photonics* where 'emerging' is taken to mean 'coming into existence', 'coming into maturity', and 'coming into prominence'. IOP Publishing and I hope that you find this series of significant value to you and your career.

Advanced Nuclear Radiation Detectors

Materials, processing, properties and applications

Ashok K Batra

Professor of Physics-Materials Science, Department of Physics, Chemistry, and Mathematics, College of Engineering, Technology and Physical Science, Alabama A&M University, 4900N Meridian Street, Huntsville, AL 35811, USA

IOP Publishing, Bristol, UK

ISBN 978-0-7503-2508-0 (ebook)
ISBN 978-0-7503-2506-6 (print)
ISBN 978-0-7503-2509-7 (myPrint)
ISBN 978-0-7503-2507-3 (mobi)

DOI 10.1088/978-0-7503-2508-0

Version: 20210901

IOP ebooks

British Library Cataloguing-in-Publication Data: A catalogue record for this book is available from the British Library.

Published by IOP Publishing, wholly owned by The Institute of Physics, London

IOP Publishing, Temple Circus, Temple Way, Bristol, BS1 6HG, UK

US Office: IOP Publishing, Inc., 190 North Independence Mall West, Suite 601, Philadelphia, PA 19106, USA

To my loving wife, Nutan

Contents

Preface

In the last decade, gamma-ray detectors have advanced at a breakneck pace. The current and upcoming materials and material-based technologies for gamma-ray detectors, as well as their growth process in various forms, such as single crystals, films, and ceramics, are reviewed in this book. The book will serve as a foundation for academic teachings as well as a guide and reference for all individuals involved in the development or application of detectors, including researchers, scientists, engineers, and business owners. As mentioned in the following paragraphs, the book is divided into five chapters.

The first chapter discusses the fundamentals of radiation interaction with matter, with a focus on gamma-ray detectors. Organic and inorganic important gamma-ray sensors are described, as well as their properties and functioning principles. Material requirements for scintillators are broken down into sections based on their use in various applications.

The main performance parameters of gamma-radiation materials are highlighted in chapter 2 along with the methodology used to determine them.

The third chapter analyzes and identifies a variety of scintillators for gamma radiation detectors, as well as their applications. Unique performance parameters. Inorganic, semiconductor, transparent ceramics, organic chemicals, organic-inorganic hybrid perovskites, and nanostructured scintillators are all discussed extensively in this chapter.

Important scintillators researched in various forms, as well as the generic fabrication methods and technologies used, are summarized in chapter 4. Bulk single crystals, polycrystalline ceramics, thin films, thick films, and composites are all part of this category. The principles of crystal development and materials processing techniques are described, along with an illustration of the methodology.

The essential uses of radiation detectors in various arenas are covered in chapter 5. The perspective and prospects for the future are also discussed. The relationship between scintillator parameters and application is summarized in table 5.1.

Foreword

I congratulate the author for having brought out timely an excellent monograph on nuclear materials for gamma-ray detectors. Gamma-ray materials and detectors remain a topic of immense interest in academic and industrial settings, including for national security as well. Dr A K Batra is a prolific scholar, passionate author, and accomplished researcher who has authored two monographs related to pyroelectric infrared sensors and ambient energy harvesting. He has contributed significantly to the advancement of knowledge in this area of inquiry and other related fields. I am sure it will also prove to be interesting, thought-provoking, and helpful in spurring further progress in this fascinating and challenging field of great importance both in the medical as well as defense industries. The style of the monograph is very clear and easy to read. The plan adopted by the author is well-conceived. Thoroughly compiled, this monograph should prove to be of immeasurable benefit and use for graduate students, engineers, technicians and scientists, managers as a comprehensive guide to current state-of-the-art science and technology of the analysis and development of gamma-ray nuclear materials for commercial, medical, industrial, military and space applications. The merit of this monograph is that it brings to both novice and advanced readers all the topics required to understand and jump-start investigation on interesting gamma-ray materials/crystals and their growth. It would be appropriate to mention that it presents a complete lifecycle of material from the basics of nuclear theory of materials, to the processing and technological applications in a systematic manner with emphasis on most recent advances. It is worth mentioning that it also contains recipes of some important materials for bulk crystal growth.

I do not doubt that this venture of the author will be a grand success, and will also encourage him in the future to author a few other advanced books.

Govindhan Dhanaraj. PhD,
VP and Chief Technologist
govindhan.dhanaraj@pallidus.com
30 Corporate Circle, Suite 101, Albany, NY 12203. Phone: 800–608–1314

Acknowledgments

First and foremost, this monograph would not be possible without the inspiration of my dear father who advised me to write this unique scientific resource. I express my gratitude to him and my late mother for instilling essential life traits in me, such as perseverance, hard work, and humility. I am forever indebted to my wife Nutan, my lifeline, for her constant prayers and her tenacity in staying by my side through thick and thin with total dedication, love, and encouragement. I am eternally grateful to my close-knit family, including dear mama and mammiji who have supported me throughout my career and in authoring this monograph. A special thanks to Professor Ashok Vaseashta for our many stimulating discussions on materials and his broad understanding of materials science. I feel privileged to have worked with eminent teachers, including Professor S C Mathur, the late Professor R B Lal, and Professor K L Chopra. I have benefited from their scientific knowledge and philosophy of life. I am grateful to Professor M D Aggarwal for introducing this subject to me. I want to express my appreciation to the university administration and the faculty and staff of the Physics department for their general support and the friendly atmosphere that they create. Special thanks to Drs Matthew Edwards, B R Reddy, Arjun Tan, and N Bhat for their encouragement and help. Additionally, I would like to acknowledge support from my research students, especially Dr M Alomari, Dr Bhir Bohara, Dr Padmaja Guggilla, Dr James R Currie, Dr Jason M Stephens, Dr Ryan Moxon, Dr Ashwith Chilvery, and Mychal Thomas. Thanks are also due to Dr Rahul Reddy for many discussions and help. Dr A Kassu deserves special mention for his technical help. Special thanks to Dr Rastgo Hawrami for many discussions and technical help.

I am also grateful to the reviewers and editor Professor Barry Johnson who gave positive comments and useful suggestions for improvement during the initial stages of the book's production.

Author biography

Ashok Batra

Ashok Batra holds a Master's of Technology and PhD from the Indian Institute of Technology, Delhi. With more than 30 years of experience in the diverse areas of solid-state physics/materials and their applications, he is currently a Professor of Physics at Alabama A&M University. His research experience and interests encompass ferroelectric, pyroelectric, and piezoelectric materials and their applications; design, fabrication, and characterization of pyroelectric, piezoelectric, photothermal and photovoltaic devices; nonlinear optical organic crystals, organic semiconductors, crystal growth from solution and melt, microgravity materials research, nanocomposites, biomedical sensors, nuclear detectors, and chemical sensors. Dr Batra is currently engaged in research related to the development of ambient energy harvesting, nanoparticle-based chemical sensors, electrospun nanocomposites, and organic photovoltaic solar cells. He has been both a principal and co-investigator for various research grant awards from the U.S. Army/SMDC, NSF, DHS, and NASA, including a NASA grant related to the International Microgravity Laboratory-1 experiment flown aboard the Space Shuttle Discovery. Recipient of a NASA Group Achievement Award and the Alabama A&M University School of Arts and Sciences Researcher of the Year award, he has published more than 200 publications, including four books, book chapters, proceedings and review articles, and NASA Technical Memorandums. Dr Batra is a member of SPIE, AES, and AAS.

IOP Publishing

Advanced Nuclear Radiation Detectors

Materials, processing, properties and applications

Ashok K Batra

Chapter 1

Interaction of gammas with matter

1.1 Introduction

Radiation has a variety of interactions with matter. The interactions between radiation and matter, namely gamma radiations, are discussed in this chapter. Charged particles such as alpha and beta particles, accelerator-created charged particle beams, and photon electromagnetic radiation are all examples of radiation.

1.2 Gamma-ray

Gamma-rays were discovered by Paul Ulrich Villard in 1900. Gamma-rays have short wavelengths and high energy from 10 keV to 5 MeV [1, 2]. Photons are particles with no charge and zero rest mass, which travel with the velocity of light. A gamma ray is emitted from unstable atomic nuclei and has high penetration power. It has low specific ionization compared to that of an alpha particle, but it is higher than a beta particle. When a monoenergetic gamma-ray passes through an absorber, a portion of the energy is transferred to the absorber in a single interaction as an exponential absorption. The amount of attenuation can be mathematically expressed as:

$$\frac{I}{Io} = \exp(-\mu t) = \exp\left(\frac{-\mu}{\rho}\right)\rho t \tag{1.1}$$

where I is the number of photons transmitted, Io is the number without the absorber, (μ) is the linear attenuation coefficient, t is the absorber thickness, and ρt is the absorber mass thickness.

The gamma-ray is created by nuclear processes and the decay of excited nuclei. The gamma-ray has tremendous penetrating power and is difficult to block because it has neither mass nor charge. Several feet of concrete or several meters of water will allow a small fraction of the original gamma-stream to flow through.

doi:10.1088/978-0-7503-2508-0ch1

1.2.1 Gamma-ray interactions with matter

For gamma-radiation, the most critical mechanisms of interaction are:

1. Photoelectric effect;
2. Compton and Rayleigh scattering; and
3. Pair production.

1.2.1.1 Photoelectric effect

The photoelectric effect is particularly essential for low-energy gamma-rays (less than 0.2 MeV). It entails an inelastic collision between a photon and an electron, in which the photon's whole energy is transferred to the electron. An electron is ejected from its orbital with kinetic energy equal to the energy of the gamma-ray that impacted it, less the binding energy that holds the electron in its orbital, in this process effectively. The photoelectric effect is schematically depicted in figure 1.1.

1.2.1.2 Compton effect

For medium-energy gamma-rays, the Compton effect (Compton scattering) is critical (0.5–1.0 MeV). It includes the collision of a photon and an electron that is somewhat elastic. A portion of the photon's energy is transferred to the electron in this process. The photon exits the collision in a different direction and with less energy. The Compton effect is depicted in figure 1.2.

1.2.1.3 Pair production

When pair production is possible, it is especially important in the presence of matter with an energy greater than 1.022 MeV. The gamma-ray fades away, and a positive and negative electron appear. The angle formed by the electrons is unknown. The pair-production process is depicted in figure 1.3. This happens mostly in the Coulomb force field, usually around an atomic nucleus, but it can also happen in the field of an atomic electron. It is worth noting that there is a reference frame in which the combined momentum of the electron and positron is zero. Photon momentum, on the other hand, is always $h\nu/c$. For momentum conservation, a third body is required: an electron or a nucleus.

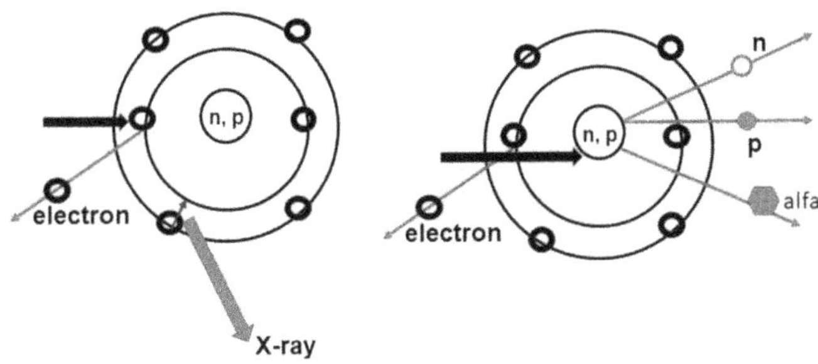

Figure 1.1. Photoelectric effect (a) and nuclear photoelectric effect (b).

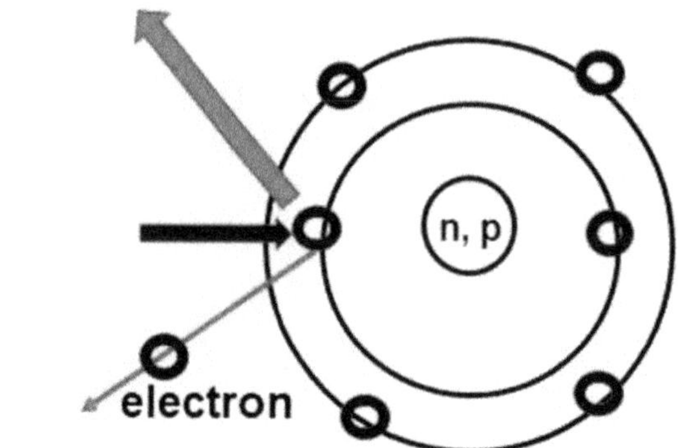

Figure 1.2. Compton scattering.

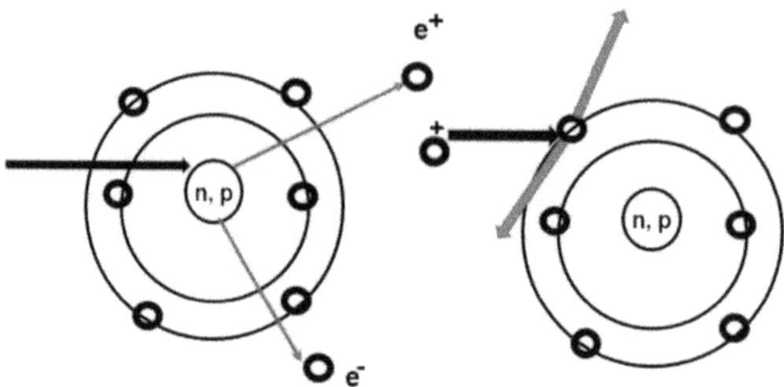

Figure 1.3. Pair production with subsequent inhalation.

1.2.2 Types of basic gamma-ray scintillators

Scintillators are categorized based on their chemical makeup. Organic and inorganic are the two major groups. Organic scintillators have a faster decay period, lower density, and are better. They can directly detect electrons and alpha particles, although they create less light. The best light output and proportionality come from inorganic scintillators, but their response time is often longer. The great density of inorganic crystals and the high Z-value of the constituents make them ideal for gamma-ray spectroscopy. Organics, on the other hand, are frequently favored. Because of the presence of hydrogen, lithium, or boron, beta spectroscopy and fast neutron detection are possible. As radiation detectors, scintillators can be used in a variety of ways. Scintillators are utilized in x-ray imaging, x-ray radiography, computed tomography (CT) scan, mammography, positron emission tomography (PET) scan, and medical diagnostics for high-resolution measurement of the spatial distribution of radiation intensity in different organs, as well as space and planetary exploration. A radiation detector is used in industrial x-ray measuring systems, exploration and mining activities, nuclear nonproliferation, homeland security, and national defense to identify radioactive sources and associated gamma-ray signals. Information about the type of radiation and its energy can be obtained by

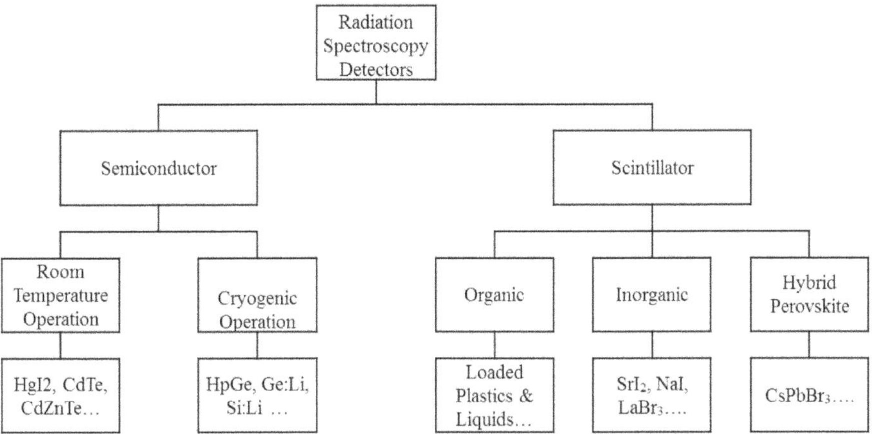

Figure 1.4. Types of radiation detectors.

identifying the isotope that produces it. The types of radiation detectors are depicted in figure 1.4.

The following are the most common scintillators:

1. Crystals of inorganic matter.
   - Alkali iodides with a low impurity content, such as NaI(Tl), CsI(Tl), CaI(Na), LiI(Eu), and CaF_2(Eu) or ZnS.
   - Excellent gamma-ray detection efficiency.
   - The pulse decay time is approximately 1 microsecond.
2. Organic scintillators are a type of organic scintillator.
   - Naphthalene, stilbene, and anthracene are aromatic chemicals.
   - A pulse decay time of approximately 10 nanoseconds.
   - Beta particle detection.
3. Scintillators made of plastic
   - Chemicals for scintillation in the plastic matrix.
   - Time it takes for a pulse to decay is 1–2 nanoseconds.
   - Detection of fast neutrons using hydrogen.
   - From thin fibers to thin sheets, it may be machined into practically any shape and size. Because they are impervious to water, air, and many chemicals, they can be utilized in direct contact with radioactive samples.
   - Gaseous scintillators are a type of gaseous scintillator. Mixtures of noble gases.
   - The scintillations are produced as a result of atomic transitions. Since the light emitted by noble gases belongs to the ultraviolet region, other gases, such as nitrogen, are added to the main gas to act as wavelength shifters. Thin layers of fluorescent materials used for coating the inner walls of the gas container achieve the same effect.

Semiconductor detectors come in a variety of shapes and sizes.

1. Oxidized high purity surface-barrier detector surface layer of Si that is p-type.
2. Detector of diffuse junctions:
   - Extremely pure p-n junction on Si wafer with n-type layer (diffused P)
3. SiLi (silicon lithium-drifted) detector:
   - Li diffusion results in the formation of an n-p junction.
   - Ion drifting increases the depth of depletion.
   - After the drifting process is finished, the detector is installed on a cryostat and kept at a low temperature at all times (77 K).
   - X-rays and charged particles.
4. Germanium lithium-drifted detector (GeLi):
   - The same SiLi preparation.
   - Maintaining a low temperature for the Ge(Li) detector is far more important than maintaining a low temperature for Si(Li) detectors, which are always operated at liquid nitrogen temperatures (77 K).
5. HPGe (hyperpure germanium) detector.
   - High-purity germanium with less than 10^{16} atoms/m^3 impurity concentration.
   - It can be either p-type or n-type.
   - Up to 200 cm^3 volume.
   - It is possible to store it at room temperature.

1.3 Principles of operation of gamma-ray detectors

1.3.1 Scintillation detectors

Scintillators are materials that emit sparks or scintillation of light when ionizing radiation passes through them. They can be solids, liquids, or gases. A scintillator was the first solid substance used as a particle detector. A scintillation counter's operation can be separated into two parts [3, 4]:

1. Scintillating material absorbs absorbed radiation energy and raises electrons to excited states. The scintillator emits a visible light photon after subsequent de-excitation of electrons.
2. A photon from the scintillating material interacts with the photocathode of a photomultiplier tube (figure 1.5), releasing electrons. With the help of an electric field, the electrons released by the photocathode are guided towards the first dynode. A material that emits secondary electrons is coated on the dynode. The secondary electrons from the first dynode are sent to the second dynode, the second dynode is directed to the third dynode, and so on. The number of dynodes in a typical commercial phototube might range from 15 to 20 [5]. The subsequent dynodes' creation of secondary electrons culminates in a final amplification of 10^6 or more. The difference in potential between two consecutive dynodes is on the order of 80–120 V.

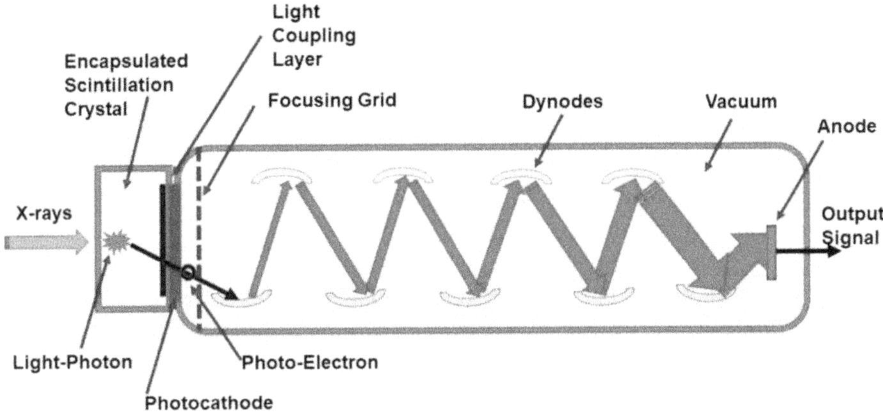

Figure 1.5. A photomultiplier tube (PMT).

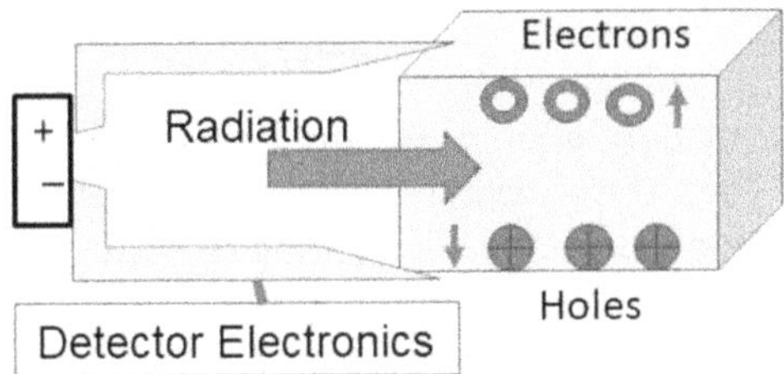

Figure 1.6. Principle of a semiconductor detector.

1.3.2 Semiconductor detectors

Operational principles: Semiconductor detectors are solid-state devices that function similarly to an ionization chamber. Electrons and holes are the charge carriers in semiconductors. As it passes through the semiconducting junction, radiation forms electron–hole pairs (figure 1.6). Under the influence of the electric field, electrons and holes flutter away. The charge collected forms a pulse that can be recorded with the right electronics.

It's important to note that when an electron jumps to the conduction band, the valence band becomes empty. This is referred to as a 'hole.' The lack of an electron is referred to as a hole. Electrons and holes travel in opposite directions. In the same way that electrons contribute to electrical transport, holes do as well. The number of electrons in an inherent and electrically neutral semiconductor is always equal to the number of holes.

The most significant advantage of semiconductor detectors over other types of radiation detectors is their superior energy resolution: the capacity to discern particle energy from a poly-energetic energy spectrum.

1.4 Material requirements for scintillators

The selection of scintillator material is based on the needs of the application. The characteristics that determine which scintillator to use are listed below [6]:

Radiation absorption efficiency: also known as absorption coefficient or absorption length, is a metric that defines how well a material absorbs radiation during the scintillation conversion step. Materials with a high density and atomic number are desirable for x-ray and gamma-ray detection.

Light yield: The most essential metric is light yield (LY), which is defined as the number of photons released per unit of deposited energy. It's helpful for determining scintillator efficiency, sensitivity, and energy resolution. The LY is determined by the number of electron–hole pairs that can be created in the ionization tracks as a result of the incident photon's interaction with the scintillating material. The amount of electron–hole pairs produced, and hence the amount of light produced is proportional to the bandgap E_g of the molecule. The light yield, expressed in photons/MeV, can be calculated using the following formula:

$$LY = 10^6 \left(\frac{SQ}{\beta E_g} \right) \tag{1.2}$$

S is the electron–hole transport efficiency to the optical center, Q is the luminescence quantum efficiency of the optical center, and is a constant, generally 2.5. Because internal scattering and re-absorption occur during the transmission of light from the scintillator to the detector, the actual light yield of a scintillator, depending on the geometry of the scintillator, may be lower than the theoretically expected amount.

Response time: After being subjected to radiation, the time it takes for the scintillator to release a UV/VIS photon. The decay period of the scintillation is the key determinant of the response time. Fast response times, and thus low decay durations, are critical in applications like computer tomography and particle detection in accelerators where timeliness is critical.

Self-absorption of light: Consider the optical transmission for the nominal thickness of the scintillation spectrum while building a scintillator. If there is too much self-absorption, the converted photons will be reabsorbed and lost by non-radiative processes.

Energy resolution: The ratio of the peak energy location in the pulse height spectrum divided by the full width at half maximum (FWHM) of the peak at particular energy in response to exciting radiation is referred to as energy resolution. This property is vital for spectrum measurements of incoming radiation, particularly in γ-ray spectroscopy, as well as the scintillator's capacity to differentiate between different radiation energies. The intrinsic resolution of a scintillator is determined by the material's non-proportional response, but faults in the scintillator, such as inhomogeneities that produce local fluctuations in light output and non-uniform reflectivity, can also affect the energy resolution.

The spectra released by the scintillator after being in activated by the radiation should match the spectrum of the photo-detector to avoid post-scintillation losses.

This is known as spectral matching, and it is more of an engineering issue than a material issue because it requires the detector's sensitivity and efficiency to be at their highest in the spectrum area where the scintillator emits. However, to match the emission of commercially available detectors, materials may need to be changed.

Stability: Chemical and radiation stability are two types of stability. Chemical stability refers to a material's inherent stability, which includes self-life. Radiation stability, also known as radiation hardness, refers to a material's capacity to resist considerable degradation when exposed to radiation. The stability of a material dictates how long it can be used as a scintillator before being replaced

Proportionality: The scintillation reaction should be proportional to the incident radiation, as this can impair intensity discrimination. In many cases, having a linear response of the scintillator, at least in the energy range of interest, is advantageous.

1.5 Detailed scintillation mechanism

As electrons are lifted from the valence band to the conduction band by a charged particle traveling through the scintillator, a significant number of electron–hole pairs are formed. Because the activator's ionization energy is lower than that of the lattice site, the holes will swiftly drift to the location of an activator site and ionize it. The electron can freely move through the crystal until it comes into contact with an ionized activator. Now that the electron has entered the activator site, it can form its own set of excited energy levels. If the newly generated activator state is an exciting configuration with an allowed transition to the ground state, de-excitation will occur swiftly with the possibility of photon emission. Lifetimes are typically in the range of 30–500 nanoseconds. Because the electron migrates at a considerably faster rate, all of the excited activator sites form at the same time and then de-excite with the excited state's half-life. As a result, the time characteristic of the emitted scintillation light is determined by the decay duration of these states. It is possible that a less enticing option exists. When an electron reaches the activator site, it forms an exciting configuration that prevents it from returning to its ground state. To raise them to a higher-lying condition that allows de-excitation to the ground state, further energy in the form of a phonon is required. Phosphorescence is the name given to the phonon energy and the consequent slow component of light. In scintillators, it is a key source of afterglow. Quenching is another unfavorable loss mechanism. which occurs when an electron's de-excitation causes non-radiative relaxation to the ground state. At various energy levels, figure 1.7 demonstrates the concepts of absorption, fluorescence, and phosphorescence.

1.5.1 Scintillation mechanism in an inorganic scintillator

Organic and inorganic scintillators have different scintillation mechanisms.

Figure 1.8 illustrated the principle of scintillation light production and the scintillation mechanism. The energy cascade, thermalization and migration of charge carriers to the luminescence center and excitation and emission of the luminescence center are the three stages of the scintillation mechanism. The energy

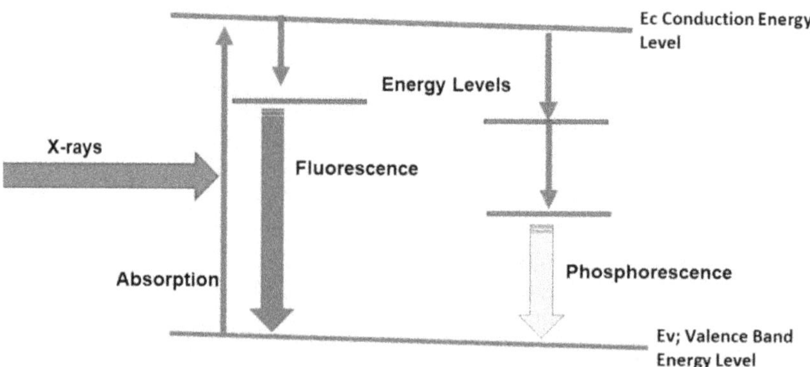

Figure 1.7. Illustration of absorptions, fluorescence, and phosphorescence.

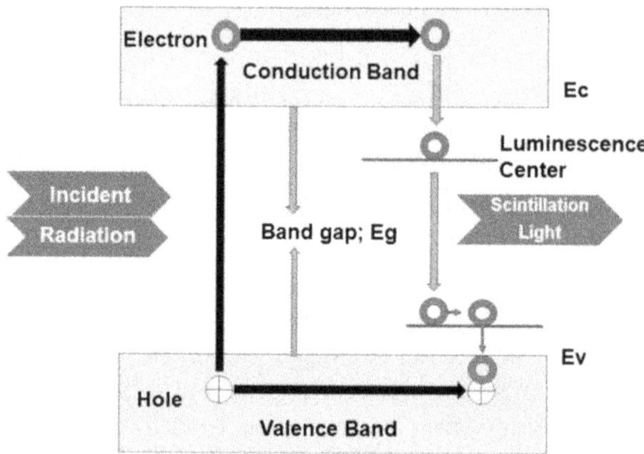

Figure 1.8. Illustration of scintillation light production and scintillation mechanism.

cascade begins with atom A absorbing a high-energy photon, $h\nu$. As indicated below, the formation of a high-energy primary electron, $e-$, and an inner shell hole:

$$A + h\nu \rightarrow A^{+} + e^{-} \tag{1.3}$$

The inner shell hole emits a photon radioactively and creates an Auger electron non-radiatively, while the primary electron relaxes in different ways. A phonon released during radiative relaxation may escape the crystal or be absorbed by another atom, resulting in a new deep hole and a liberated electron. The likelihood of nonradiative decay is usually larger than that of radiative decay [3]. The Auger electron loses energy due to electron scattering. Atomic inelastic scattering on electrons (electron–electron relaxation), the primary electron relaxes as follows:

$$A + e^{-} \rightarrow A^{-} + 2e^{-} \tag{1.4}$$

After ionization, the initial electron that generated ionization and the manufactured secondary electrons are indistinguishable, The secondary electron created may trigger more ionization, resulting in an avalanche of electrons and holes. [4]. Until the energy of the electrons falls below $2E_g$, the ionization threshold, this electron multiplication process continues. In most circumstances, Beyond the auger process threshold, electrons develop in the conduction band, while all holes form in the valence band, with no core band formed. When primary electrons interact with valence electrons, they lose their energy and form plasmons. The number of electron–hole pairs, N_{eh}, formed when a high-energy photon is absorbed is provided by the following ratio:

$$N_{eh} = \frac{E_\gamma}{\beta E_g} \tag{1.5}$$

where E_γ is the photon energy, E_g is the bandgap, and β is a constant (1.5–2 for ionic crystal scintillators).

Charge carriers thermalize and migrate to the luminescence center when heated electrons and holes interact with the host lattice through electron–phonon relaxation near the end of the energy cascade. On a time scale of 10^{-14} to 10^{-10}, the second step takes place. The holes move to the top of the valence band as the electrons move to the bottom of the conduction band. For ionic crystal scintillators, the thermalization process is defined as carrier migration with a characteristic length $L = 10^2$–10^3 nm. Electrons and holes can be collected by traps during migration, resulting in a self-trapped exciton (STE) that is both free and bound.

The charge carriers recombine at $t > 10^{-9}$ at the end of the thermalization phase to create intrinsic or extrinsic luminescence. Excitonic and cross luminescence are the two forms of intrinsic luminescence. The excitonic luminescence caused by the radiative annihilation of STEs has been observed in a variety of inorganic scintillators [4]. The formation of STEs in inorganic scintillators is examined using the SrI_2 scintillator as an example of thermalization and migration. Charge carriers are delivered to the luminescence center. Nonradiative processes take control at high temperatures, hence excitonic luminescence is only effective at low temperatures. Cross luminescence, also known as core-to-valence luminescence, is another type of intrinsic luminescence that involves valence band electrons and the outermost core band. Extrinsic luminescence is commonly caused by rare-earth ions in the bandgap of the host matrix. Rare-earth ions are excited by impact excitation, which is followed by electron–hole capture. Figure 1.9 depicts the potential energy diagram of a luminescence center. The configuration coordinate, also known as the distance between a luminescence center and the surrounding ions, is shown versus the energy. The ground and excited states of the center are represented by points A and B, respectively. A non-equilibrium energy state is formed when a transition from A to C occurs during excitation, and the system relaxes from C to B, releasing a phonon. A phonon is released during the transition from B to D, and another for the transfer from D to A. Depending on the electron density at B, excited electrons can be thermally accelerated to reach F, where they can transition from the excited state to the ground state via lattice vibrations (quenching). A Stokes shift is also possible, in which the absorption band A and emission band B change along the x-axis.

Dopants, also known as activators, increase the chance of photon emission from de-excitation of luminescence centers. Thallium (Tl^{1+}), sodium (Na^{1+}), cerium (Ce^{3+}), and europium (Eu^{2+}) are some of the most regularly utilized activators.

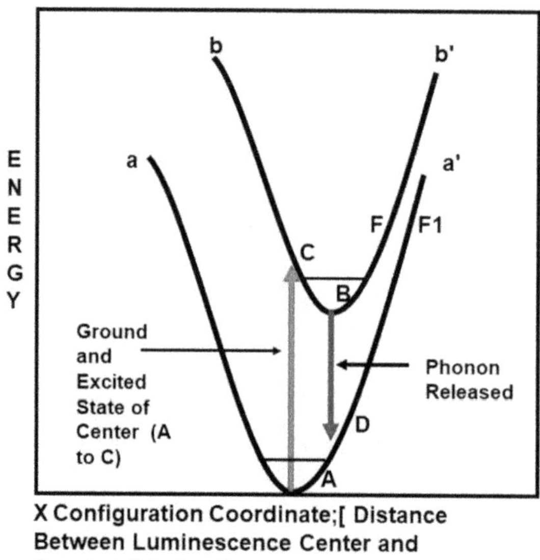

Figure 1.9. Potential energy diagram of a luminescence center.

The following are the abbreviated names for doped scintillators: Na:Tl^{1+}, SrI_2:Eu^{2+}, CsI:Tl^{1+} (thallium doped cesium iodide), CsI:Tl^{1+} (thallium doped cesium iodide). The addition of doping results in interstitial bands in the material's bandgap. It is feasible to induce scintillation at visible wavelengths by carefully selecting the doping element. Ionizing radiation is converted into optical photons at a high rate. It is referred to as high light output. A high light output or light yield is defined as more than 25 000 photons/MeV. The unit is the number of photons emitted per unit of energy received, which is usually 1 MeV.

$$L = \frac{E_\gamma}{\beta E_{\text{Bandgap}}}(S \times Q) \tag{1.6}$$

The theoretical light yield of inorganic scintillators can be calculated as follows if the efficiency of charge carrier transfer to the luminescence center, S, and the quantum efficiency of the luminescence center, Q, are both assumed to be 100%, $S = Q = 1$, and the energy required to create an electron–hole pair is assumed to be equal to $2E_{\text{Bandgap}}$.

$$L = \frac{0.5}{E_{\text{Bandgap}}} \times 10^6 \left[\frac{\text{photons}}{\text{MeV}}\right] \tag{1.7}$$

This means that the bandgap has a big impact on the light yield of inorganic scintillators. A scintillator's energy resolution refers to its capacity to distinguish between two photo peaks that are near together. The 662 keV photopeak of a ^{137}Cs source has a good resolution of less than 3% when the complete FWHM is measured. The following is a definition of resolution:

$$R\Delta\frac{\Delta E}{E}\times 100\,[100\%] \tag{1.8}$$

where ΔE is the FWHM of the photopeak located at energy E.

The time it takes for the emission intensity to drop to e^{-1} is known as the scintillation decay period. In scintillators, the rise time of emission intensity is usually less than the decay time, and it is considered to be at $t = 0$. Most scintillators have a basic exponential decay function that describes their decay duration:

$$I_t = I_o\exp\left(\frac{t}{\tau}\right)\quad [s] \tag{1.9}$$

where I_t is the number of photons emitted at time t, I_o denotes the total number of photons emitted at time $t = 0$, and is the decay constant. Multiple decay processes are observed in some scintillators, for which a more sophisticated component exponential decay function is the best way to describe them:

$$I_t = \Sigma I_n\ \exp\left(\frac{t}{\tau_f}\right) \tag{1.10}$$

Throughout these many decay processes, a fast, τ_f, and a slow, τ_s component are frequently observed. Fluorescence requires a quick (10 ns–5 μs) scintillation decay period, depending on the desired count rate in applications, to minimize instrumental signal pile-up. The emission of light by a substance that has absorbed light or other electromagnetic energy is known as fluorescence. Radiative transitions from the lowest excited singlet levels to the ground state with a typical millisecond lifespan produce this light. When the decay rate is too slow, phosphorescence occurs, resulting in afterglow. A non-radiative initial conversion converts the excited state to a metastable triplet state. Phosphorescence is caused by a metastable triplet state with a non-radiative initial conversion from the excited state. Below the lowest excited singlet states is the metastable state. Over time, the metastable electron may accumulate enough thermal activation energy to revert to the lowest excited state, resulting in delayed fluorescence. Afterglow is the light that remains after the primary decay period of the main emission has passed; the afterglow decay time, τ_s, is given by:

$$I_t = \left(\frac{kN_s}{\tau_s}\right)\exp\left(\frac{t}{\tau_s}\right) \tag{1.11}$$

The likelihood that a luminous center would capture the carrier is k, and the number of carriers trapped by traps is N_s.

A virtuous scintillator should be radiation resistant. Radiation damage that causes dislocations and shallow traps should be avoided by the detector. Dislocations within the matrix cause optical centers to develop, which can absorb transmitted photons, resulting in reduced transmission of light. The amount of afterglow produced by radiation-induced shallow traps increases, lengthening the decay time. A high density is required for a good scintillator. For gamma-ray and high-energy particle detection, high-density scintillators are preferred. A high atomic number, Z, improves gamma quanta absorption via a photoelectric effect. The gamma-ray absorption per cubic centimeter of the photoelectric effect is proportional to the product of the material's density and the effective atomic number cubed [7].

A secondary detecting medium, such as a PMT, photodiode, or other photodetectors, is required when using a scintillator. As a result, the importance of emission in the photodetector spectrum response cannot be overstated. A scintillator's emission wavelength should be close to the photodetector's maximal quantum efficiency. The light output of the scintillator must not only be within the spectral response of the PMT, but it must also have a refractive index that is close to that of the PMT window. The light is fully internally reflected if the refractive index, n, of a scintillator is internally reflected beyond the critical angle (crit), a proportion of the light intensity incident on a scintillator–air interface under normal conditions

$$\Theta_{\text{crit}} = \sin^{-1}\left(\frac{1}{\mu}\right) \tag{1.12}$$

Solid scintillators have a critical angle of less than 45 degrees because their refractive index is greater than the square root of $2^{1/2}$. The primary goal of many radiation detector applications is to determine the energy distribution of the incident radiation. This can be tested by observing how it reacts to a single source of that radiation. A radioisotope of cesium, ^{137}Cs, with a photon energy of 662 keV, is a popular monoenergetic gamma-ray source. The response function is the energy response of the detector when it is exposed to a radioisotope source, and it is quantified by the FWHM of the peak at the pulse height. Time response is a basic property of scintillators as mentioned above it represents the decay time of prompt fluorescence. An approach to describing these phenomena assumes that the population of the optical levels is also exponential so the overall shape of the light pulse is given by:

$$I = I_o(e^{\frac{-t}{\tau}} - e^{\frac{-t}{\tau_1}}) \tag{1.13}$$

where τ_1 is the time constant that describes the optical levels' population, and is the time constant that describes their decline. A Gaussian function $f(t)$ with a standard deviation σ_{ET} is frequently a better representation of the population step. After that, the whole light versus time profile is summarized as follows:

$$\frac{I}{I_o} = f(t)\, e^{\frac{-t}{\tau}} \tag{1.14}$$

The FWHM of the resulting light versus time profile can be used to characterize. In experiments, the light output rises and falls. A scintillator's light output is another important aspect. In a scintillator, only a small percentage of the energy lost by a charged particle is converted into visible light. The efficiency of scintillation is a fraction of particle energy that is affected by particle type and energy. A resolution with an FWHM of less than 4% is regarded as satisfactory. The squared total of three contributions to energy resolution is defined as R.:

$$R^2 = R_{\text{stat}}^2 + R_{\text{inhom}}^2 + R_{\text{nonprop}}^2 \tag{1.15}$$

where R_{stat} drops as the number of detected photons increases, R_{nonprop} is the contribution owing to nonproportional response, and R_{inhom} is the contribution due to inhomogeneous qualities of the crystal, reflector, photocathode, and so on. The variable in the number of detected photons, N_{dp}, plus the variance υ (= 0.1–0.2) owing to electron multiplication in the PMT account for the statistical contribution, R_{stat}. Poisson statistics are used to describe this contribution:

$$R_{\text{stat}} = sqr\left(1 + \frac{\nu}{N_{dp}}\right) \tag{1.16}$$

1.5.2 Prefered properties of scintillators

The application of scintillators can be tweaked to improve certain qualities. A good scintillator has high light output, high energy resolution, quick scintillation decay time, resistance to radiation damage, high density, and emission in the photodetector spectral response.

References

[1] Knoll G F 2010 *Radiation and Measurement* (New York: Wiley)

[2] Kapoor S S and Ramamurthy V S 1986 *Nuclear Radiation Detectors* (New Delhi: NAIL, Publishers)

[3] Rodnyi P A 1997 *Physical Process in Inorganic Scintillators* (Boca Raton, FL: CRC Press LLC)

[4] Rowe E 2014 High-performance doped strontium iodide crystal growth using the modified Bridgman method *PhD Thesis* Virginia Commonwealth University, VA https://scholarscompass.vcu.edu/etd/3397

[5] www.Hamamatsu.com

[6] Maddalena F *et al* 2019 Inorganic, organic, and perovskite halides with nanotechnology for high-light yield X- and γ-ray scintillators *Crystals* **9** 88

[7] van Eijk C W 2001 Inorganic-scintillator development *Nuc Instr. Meth. Phys. Res.* A **460** 1–14

Chapter 2

Performance of gamma radiation detectors materials

The performance of a radiation detector material must be evaluated in the context of the application for which it will be used in the manufacture of a detector system. In the following section, the performance metrics for national security and medical diagnostics, in brief, are described. The performance parameters considered are energy resolution, rate and timing, spatial resolution, detector efficiency, geometric efficiency, and operational factors.

2.1 The performance parameters [1]

2.1.1 Energy resolution

The identification of the radioactive isotope or isotopes responsible for the radiation is one of the most important requirements for gamma-radiation detection. This information can be used to determine whether a product contains naturally radioactive isotopes, medically or economically beneficial radioisotopes, or possibly dangerous radioisotopes. High-resolution energy spectroscopy is essential because each radioisotope emits a unique distribution of gamma radiation energies. As a result, the energy resolution of gamma radiation detectors is an important performance indicator for national security applications. The quantity, configuration, and type of radionuclides present can be determined using a gamma-ray detector with sufficient energy resolution.

The full-width half-maximum (FWHM) of the peak that would be recorded when that material is employed to capture a gamma-ray energy spectrum is typically used to describe the energy resolution of a radiation-detection material. While energy resolution is a function of energy, the benchmark gamma-ray of ^{137}Cs at 662 keV is widely used as an example in the field of radiation detection. The energy resolution of a radiation-detection material is influenced by a number of factors. A radiation-detection substance that delivers the highest mean number of information carriers

doi:10.1088/978-0-7503-2508-0ch2

with the least random variance between occurrences, on the other hand, is frequently undesired. Non-proportional (or nonlinear) response to the energy is one key cause for this difference in scintillators, but not in semiconductors.

2.1.2 Rate and timing

The quick detection of radiation interactions is required for a range of applications. . This can happen when describing a fairly strong radiation source, background radiation levels are high, or active technology that requires the utilization of a strong radiation source for interrogation is employed. Because background gamma-ray flux levels frequently surpass 10^3 m^{-2} s^{-1}, and correct event classification dictates that discrete events only rarely coincide, a radiation-detection material that allows rapid reading of information carriers is necessary. This necessitates strong electron and hole mobility in semiconductors and rapid and complete scintillation in scintillators.

2.1.3 Spatial resolution

In imaging applications such as radioactive source localization and medical diagnostics, the position of radiation interaction in the detector material must be precisely established. The incidence direction of the interacting radiation must also be specified in order to differentiate signals from background radiation in national security applications. Although there are other technological techniques to achieving these objectives, they all include assessing the timing and quantity of signals collected at various locations on the radiation detector material's exterior.

2.1.4 Efficiency of detection

A radiation detector's detection efficiency determines how much of the incident radiation quanta reaches the detector and is detected. A closely related metric, full-energy peak efficiency, indicates the fraction of detection occurrences in which the incident radiation's full energy is deposited. Material properties determine this efficiency in three ways. First, as the dimensions of the active medium (thickness and cross-section) rise, so do the efficiencies. Second, both efficiencies rise as the mass density of the active medium increases. Finally, the active medium's composition affects both efficiencies, which rise as the atomic number, Z, of the material elements rises. The creation of radiation-detector materials with higher density and Z, as many of the first benchmark materials were low density, has been a major focus of research for many years. Most gamma-energy spectroscopy materials must have large cross-sectional areas (10 cm^2 or more) and thicknesses of 1 to 10 cm in order to obtain satisfactory detection and full-energy peak efficiencies for gamma rays of interest. Many promising radiation-detection materials have struggled to achieve these sizes in high-purity, defect-free single crystals.

2.1.5 Geometric efficiency

The geometric efficiency simply refers to the proportion of a source's radiation quanta that strike an active radiation-detection medium. The active medium's efficiency will logically scale with the cross-sectional area it provides to the source. Due to the standoff distances necessary, many security applications, such as portal monitors, require geometric efficiency that can only be achieved with quite large detection media, maybe on the size of 1 m^2. This clearly necessitates the development of radiation-detection materials that can be produced at a reasonable scale.

2.1.6 Detection of neutron

For a neutron reaction that is favorable for neutron detection, a large cross-section is desired to increase the chances of detection. 3He, 6Li, 10B, and Cd and Gd isotopes are the most commonly employed nuclides for this purpose. However, simply detecting neutron interactions is insufficient; it is also necessary to distinguish neutron interactions from the far higher number of gamma-ray interactions that occur at the same time. Differences in the nature of the ionizing radiation produced during interaction, the energy released during interaction, and the speed of the incident radiation must all be exploited in order to achieve this. To slow down the neutron and improve detection efficiency, many neutron detectors incorporate an external hydrogenous moderator. However, neutron spectrometry, which necessitates the fast neutrons at the start, is ruled out.

2.1.7 Operational factors

In fields such as fundamental science or medical diagnosis, security applications have more strict operational requirements than other radiation-detection applications. Cost and the simplicity with which detecting equipment may be built from a certain material are operational factors. Less appealing are materials that require freezing or cryogens, are hygroscopic, or are sensitive to temperature changes or mechanical stress.

2.2 Characterization of the scintillator crystal

The cesium 137 isotope is commonly used to describe detectors based on scintillators (^{137}Cs). ^{137}Cs decays to barium 137 isotope via beta particle emission (^{137}Ba). In 94 percent of the time, the ^{137}Ba isotope will achieve a ground state by emitting a 662 keV gamma-ray, whereas in the remaining 6%, a conversion electron will be emitted followed by a 32 keV K-shell characteristic x-ray.

2.2.1 Energy resolution and light yield

The cesium-137 isotope is commonly used to describe scintillator-based detectors. Cesium is a gleaming gold-colored metal with a melting point of 28.4 °C that reacts aggressively with oxygen or water. Cesium-133 is the only stable and naturally occurring non-radioactive isotope among the forty known radioactive isotopes. These isotopes have atomic weights ranging from 112 to 151. Human action produces all other isotopes, including the more common cesium-137 (^{137}Cs), which is created by reactors as a fission byproduct. Before degrading to non-radioactive barium with a half-life of 30.17 years, ^{137}Cs produces beta particles and gamma rays as it decays to barium-137. The unstable ^{137}Ba would achieve a ground state 94% of the time by emitting a 662 keV gamma-ray, with the remaining 6% emitting a conversion electron followed by a 32 keV K-shell characteristic x-ray.

The photopeak is calculated at half the maximum height of the photopeak when it is quantified using energy resolution. The variation in the number of ionizations and excitations created causes the photopeak width. The scintillator reaction to specific radiation is measured by light yield. The observable scintillator light production is greatly influenced by reflector type, optical coupling to the photomultiplier tube (PMT), crystal geometry, and photon transfer efficiency to the PMT's photocathode. Absolute and relative light yields are the two forms of light yields. E_{ph} is the ratio of total photon energy to the energy deposited in a scintillator. η is known as absolute light yield. E_{γ}:

$$\eta = \frac{E_{ph}}{E_{\gamma}} \tag{2.1}$$

Absolute light yield is difficult to determine due to the difficulties and errors associated with evaluating scintillators such as light-gathering efficiency. The ratio of the total energy of scintillation photons T_{ph} measured at the photocathode to ionizing radiation-deposited energy E in the scintillator is known as relative light yield:

$$L = \frac{T_{ph}}{E_{\gamma}} \tag{2.2}$$

The relative light yield, which is given in this and other papers, is computed using one of four methods such as pulse, single electron, intrinsic resolution, and comparison. To calculate the scintillation photons, the signal amplitude of the PMT after the preamp is measured in the pulse method technique:

$$T_{ph} = \frac{V\varepsilon C}{\Sigma \mu E e} \tag{2.3}$$

where V is the signal amplitude measured at the preamp, T_{ph} is the photon energy, C is the PMT output capacity, the PMT anode sensitivity (PMTAS), E is the incident gamma radiation energy (662 keV for a ^{137}Cs source), and e is the electron charge. The observed single amplitude is compared to the single electron distribution of the PMT using the single electron method. The intrinsic resolution approach of the PMT calculates the number of photons N_p by counting the number of photoelectrons N_pe created at the photocathode:

$$N_p = \frac{N_p e}{Q_{\text{PMT}}} \tag{2.4}$$

The quantum efficiency of the PMT is denoted by Q_{PMT}. The comparison approach is the most extensively used of the methods mentioned above. The signal amplitude after quantum efficiency correction is compared to that of a recognized scintillator, usually sodium iodide or bismuth germanate (BGO).

2.2.2 Decay time

The fast component is mostly generated by excitonic emission, whereas the slow component is primarily caused by sequential electron–hole or hole–electron capture, as well as excitation of certain metastable states that result in emission. To determine decay time constants, single or multiple exponential decay functions fitted to the experimental data are utilized.

Not just for scintillators, but also for cathodo-luminophores, x-ray screens, and non-optically pumped lasers, the process of energy dissipation in solids is critical. In all condensed matter, the first three steps are the same. The physical mechanism of scintillation is divided into two components for ease of understanding: (1) the formation of e–h pairs (the first three steps); and (2) the excitation and emission of luminescence centers. The light output of a scintillator is proportional to the energy deposited in a crystal, according to a basic principle of pulse height spectroscopy. Combining a scintillator with a photomultiplier tube is the usual method for detecting scintillation light (PMT). A gamma-ray spectrometer also often includes a preamplifier, the main amplifier, and a multichannel analyzer (MCA) [2]. The electronics amplify the PMT charge pulse to produce a voltage pulse appropriate for detection, which is then analyzed using the MCA. The electronics intensify the PMT charge pulse, producing a voltage pulse that the MCA can detect and analyze. Figure 2.1 depicts a schematic representation of the experimental setup. Figure 2.2 depicts a typical pulse height spectrum.

A PMT is a device that converts light pulses into electrical signals. The PMT is a multipurpose light-sensitive device in use today that gives extremely high sensitivity and rapid response. In a vacuum tube, a typical PMT has a photoemissive cathode (photocathode), focusing electrodes, an electron multiplier, and an electron collector (anode). When light passes through the photocathode, photoelectrons are emitted into the vacuum. The focusing electrode-voltages then route these photoelectrons to the electron multiplier, where they are multiplied via the secondary emission process. The anode collects the multiplied electrons as an output signal. PMTs give extraordinarily high resolution due to the growth of secondary emissions The photosensitive sensors currently utilized to detect radiant radiation in the ultraviolet,

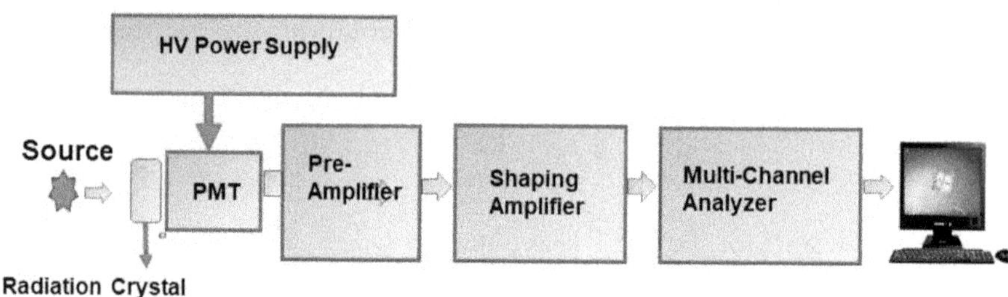

Figure 2.1. The schematic diagram of the experimental setup of the scintillation system.

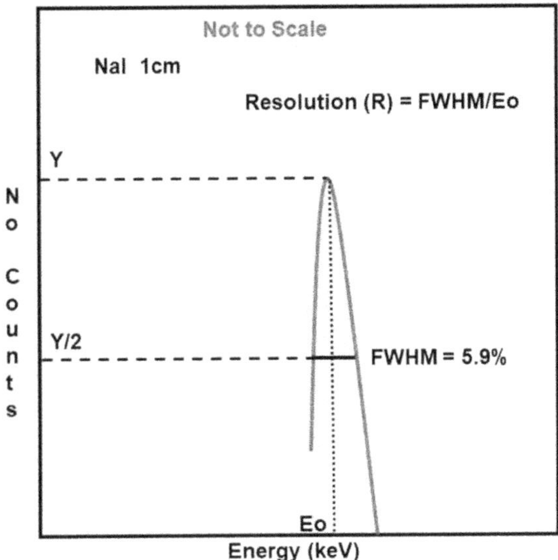

Figure 2.2. The energy spectrum of 661.6 keV gamma rays from a ^{137}Cs measured with NaI crystal.

visible, and near-infrared ranges have excellent sensitivity and extraordinarily low noise. The PMT also has a quick temporal response, minimal noise, and a huge photosensitive region to choose from.

Under ionizing radiation, pulse height measurements are commonly used to gain information on the light production, energy resolution, and decay time of the scintillating material. Figure 2.1 shows a schematic representation of a common setup for capturing pulse height spectra. With silicon-based grease, the crystal is optically linked to the PMT window (Viscasil 60 000 cSt, General Electric). Due to total internal reflection, if there is air between the scintillator and the PMT, the light beam pointed to the scintillator surface at angle arcsin (1/nSc) will not be able to leave the scintillator (nSc is the refractive index of the scintillator). It is wrapped in numerous layers of 0.1 mm thick UV reflecting Teflon tape to collect as much light as feasible. The crystal is excited by gamma and/or x-rays emitted by a radioactive source. In most cases, a ^{137}Cs source with a 662 KeV output is used.

A Hamamatsu R1791 PMT detects the scintillation light (P). The output signal is merged with a charge-sensitive preamplifier with a 50-second RC-time. An Ortec 572/672 spectroscopic amplifier then amplifies the signal again and filters it using Gaussian shaping. Shaping times range from 0.5 to 10 seconds. The time it takes for the output of the preamplifier to be integrated and shaped into a Gaussian pulse is referred to as shaping time. Finally, a standard ADC unit analog-to-digital converter converts the analog signal to a digital pulse (ADC). A computerized data handling system can then store and process the signal (PC).

References

[1] Milbrath B D, Peurrung A J, Bliss M and Weber W J 2008 Radiation detectors materials: an overview *J. Mater. Res.* **23** 2561–81

[2] Hawrami R 2008 *PhD Dissertation* Alabama A&M University, Alabama

IOP Publishing

Advanced Nuclear Radiation Detectors

Materials, processing, properties and applications

Ashok K Batra

Chapter 3

Advanced materials for gamma radiation detectors

3.1 Introduction

In 1885, scintillating materials research began. When pioneers Roentgen and Edison realized that $Ba[Pt(CN)_4]$ emitted visible scintillating photons when irradiated with x-rays, they discovered ionizing radiation. With the discovery of x-rays and photo-detectors, many types of scintillators advanced [1]. Alkali halide materials such as NaI and CsI, as well as natural phosphors (e.g., CaF_2 and $CdWO_4$), were investigated at the start of the scintillator's R&D history [2–5]. It's worth noting that in mineralogy, CaF_2 and $CdWO_4$ are known as fluorite and scheelite, respectively. In the 1940s and 1960s, well-known scintillators such as Tl-doped NaI, Tl-doped CsI, and Eu-doped CaF_2 were produced [2–5]. After the 1980s, new materials were discovered that did not occur in nature, such as $Bi_4Ge_3O_{12}$ (BGO) [6]. Following the development of BGO, significant R&D was carried out, the most prominent of which was Ce-doping in insulating host materials, which has since become a standard strategy for developing new scintillators. Ce-doped Lu_2SiO_5 (LSO) and its Y-admixture LYSO have become commonplace [7, 8]. $Gd_3(Al, Ga)_5O_{12}$ (GAGG) with Ce doping was created recently and is currently commonly utilized with Si-based photodetectors. Yanagida *et al* were involved in the development of GYAG, GAGG, Eu:SrI_2, and $LiCaAlF_6$ [9–14]. The combination of an insulator host with a doped emission center has become a typical technique to attain improved performance in the recent trend of building host materials of complicated compositions.

3.2 Inorganic scintillators

Scintillators in x- and γ-ray detectors must have a high-density ρ (in g cm^{-3}) and a low Z_{eff}. Photoelectric absorption (also known as photo-absorption), Compton scattering, and pair formation are the three major types of x- and γ-ray interactions

doi:10.1088/978-0-7503-2508-0ch3

with matter, and their interaction probabilities are determined by $Z_{eff}4$, Z_{eff}, and $Z_{eff}2$, respectively. Photoelectric absorption events are frequently used in practical detectors (pulse-counting type) because they represent information about the incident x- or γ-ray energy. As previously stated, the initial scintillators were simple chemical compounds like NaI and CsI, as well as a natural phosphor. $CdWO_4$ and CaF_2 are examples of such materials. The BGO scintillator, which has been around since the 1970s, was the second-generation scintillator. Since 1990, third-generation scintillators have been used, which are often a combination of host and emission center type (e.g., Ce-doped LSO and LYSO). As a result, the majority of today's state-of-the-art scintillators have an emission center doped in them. Rare-earth silicate, rare-earth garnet, and certain halide compounds are three trends in third-generation scintillators. Hitachi Chemical, Japan [15], discovered Ce-doped Gd_2SiO_5 (GSO) as the root of rare earth silicate. Lu-substituted scintillators, such as LSO and LYSO, were created in different parts of the world as a result of this work. The Ce-doped silicates' scintillation capabilities are characterized by a high light yield (10 000–30 000 ph MeV^{-1}). They exhibit a short scintillation decay (30–60 ns), with blue emission at 430 nm. These characteristics are ideal for positron emission tomgraphy (PET) detectors. Most PET uses LSO or LYSO because Lu-based silicate scintillators have a higher light yield (20 000–30 000 ph MeV^{-1}) than GSO (10 000 ph MeV^{-1}). Ca can be added as a co-dopant to increase scintillation characteristics, which is presently employed in practical detectors [16, 17]. Ce-doped $Gd_2Si_7O_7$ (GPS), a pyrosilicate scintillator developed by Hokkaido University and Hitachi Chemical, was claimed to have a larger light production than the LSO series and quick decay of many 10 ns [18, 19].

Later, La admixed GPS (LGPS) was discovered, which has similar scintillation behaviors to Ce-doped GPS, but is easier to produce in single crystal form [20]. The Ce-doped YAG has primarily been employed in charged-particle detectors, and it has been attempted to replace the Y with Lu (Ce doped LuAG) in order to improve detection efficiency for high-energy photons [21–24]. Gd substitution was studied instead of Lu, and Ce-doped $(Gd,Y)_3Al_5O_{12}$ (GYAG) was first developed [25]. Following further research, Ce-doped $Gd_3(Al,Ga)_5O_{12}$ (GAGG) was discovered to have a characteristic scintillation light yield (946 000 ph MeV^{-1}) and became a commercial product. Researchers developed a Ce-doped GAGG ceramic a few years ago that had the highest light yield among the oxide scintillators, achieving 70 000 ph MeV^{-1} [26]. They have continued to study Ce-doped garnets, and they have just discovered that Ce-doped $Tb_3Al_5O_{12}$ (TAG) has highly appealing scintillation qualities, with a high light yield of 57 000 ph MeV^{-1} and a decay period of 38 ns. Ce-doped garnet scintillators are the only option for Si-PD, avalanche PD (APD), and Geiger-mode APD at the moment. Pr_3D doped LuAG also demonstrated a Pr_3D 5d-4f transition in the 2000s [27]. As mentioned above, the history of scintillators and scintillation detectors that started with halide NaI, CsI, and CaF_2 are examples of scintillators, and NaI has become a standard scintillator. Continuous efforts have been undertaken to produce a 'next' NaI, because halide scintillators have a greater potential to demonstrate a high light yield due to their lower bandgap than typical materials such as oxide, fluoride, and nitride. Ce-doped

$LaCl_3$ and $LaBr_3$ have been discovered, and they have a high light yield (960 000 ph MeV^{-1}) and a high energy resolution (3%–4% at 662 keV). The largest light yield of scintillators for γ-ray detectors was at most 40 000–50 000 ph MeV^{-1} before the discovery of these materials, and this finding was highly significant.

Following the pioneering discoveries mentioned above, a number of researchers began looking at a variety of halide compounds as potential scintillators, and some novel materials were discovered. Eu-doped SrI_2 [28] was one of these newly invented scintillators that became a commercial product for gamma-ray detectors. The crystal quality had increased, and the scintillation qualities had greatly improved (e.g., scintillation light production of up to 120 000 ph MeV^{-1}, energy resolution of 3% at 662 keV, and a decay period of roughly 1 μs), thanks to recent advances in crystal growing technology.

Researchers have recently begun looking for non-hygroscopic halide scintillators. Cs_2HfCl_6 [29, 30] has emerged as a viable option in the last several years. Because Cs_2HfCl_6 is non-hygroscopic and contains Hf as a host constituent, it has a large cross-section for high-energy photons. Under gamma-ray irradiation with a few seconds decay period, the light production was estimated to be roughly 30 000 ph MeV^{-1}. Tl- and In-doped CsCl, $TlCdCl_3$, Tl and Ce-doped Cs_2HfCl_6, Rb_2HfCl_6, Ce-doped $CsCaCl_3$, and $TlMgCl_3$ were among the non-hygroscopic halide scintillators developed further. $TlMgCl_3$, in particular, has the most promising qualities of a high light yield (46 000 ph MeV^{-1}), a high energy resolution (5% at 662 keV), and a reasonably fast decay (60 ns) among these newly created scintillators [31–36]. Traditional inorganic scintillators for gamma-ray detectors are summarized in table 3.1 [37]. The density (ρ), effective atomic number (Z_{eff}), emission wavelength (λ), and major scintillating decay time (τ) are all abbreviated in the table heading (tables 3.2 and 3.3).

Table 3.1. Summary of traditional inorganic scintillators for gamma-ray detectors. Reprinted from [37] with permission of MDPI (CC-BY SA 4.0).

Scintillator	ρ (g cm^{-3})	Z_{eff} Light yield (photons MeV^{-1})	Energy resolution (%, at 662 keV)	λ (nm)	τ (ns)
NaI:Tl	3.7	50.8 43 000	6.7	415	230
CsI:Tl	4.5	54.0 66 000	6.6	560	1000
CsI:Na	4.5	54.0 43 000	7.4	425	
CsF	4.6	53.2 1900	19	390	2–4
BaF_2	4.9	52.7 1430	10	175	0.8
$CaWO_4$	6.1	75.6 15 800	6.3	425	6800
$PbWO_4$	8.3	75.6 140		~475	~10
$CdWO_4$	7.9	64.2 19 700	6.5	495	104
$Bi_4Ge_3O_{12}$	7.1	75.2 8200	27	505	300
$CdS:Te^{2+}$	4.8	48.0 17 000	14	640	270–3000
ZnS:Ag	4.1	27.4 73 000		450	105

Table 3.2. Summary of parameters of Ce^{3+}, Pr^{3+} and Nd^{3+} doped scintillators. Reprinted from [37] with permission of MDPI (CC-BY SA 4.0).

Scintillator	ρ (g cm^{-3})	$Z_{ef\ f}$	Light yield (photons MeV^{-1})	Energy resolution (%, at 662 keV)	λ (nm)	τ (ns)
LaF_3:Ce^{3+}	5.9	40.3	2200		290	3 and 27
LaF_3:Nd^{3+}	5.9	40.3	2000		173	6
CeF_3	6.2	41.1	4500		330	28
LuF_3:Ce^{3+}	8.3	50.2	8000		310	28
K_2YF_5:Pr^{3+}	3.1	23.9	6900		240	20
$LaCl_3$:Ce^{3+}	3.9	49.5	49 000	3.3	330	25
$PrCl_3$:Ce^{3+}	4.0	51.5	21 000	8.4	340	17
$CeCl_3$	3.9	50.4	28 000		360	25
$Cs_2LiLaCl_6$:Ce^{3+}	3.3	41.4	35 000	3.4	400	1 and 40
Cs_2LiYCl_6:Ce^{3+}	3.3	38.1	21 000	6.0	376	1 and 35
Cs_2LiYCl_6:Pr^{3+}	3.3	38.1	10 000	15.0	315	1 and 35
K_2LaCl_5:Ce^{3+}	2.9	44.1	30 000	5.1	344	1000
$RbGd_2Cl_7$:Ce^{3+}	3.7	53.9	40 000	5.0	370	90
$LaBr_3$:Ce^{3+}	5.1	46.9	67 000	2.8	358	15
$LaBr_3$:Pr^{3+}	5.1	46.9	75 000	3.2	450 to 900	11 000
$PrBr_3$:Ce^{3+}	5.3	48.3	21 000	5.5	365	6
$GdBr_3$:Ce^{3+}	4.6	52.4	44 000		350	20
Cs_3LaBr_6:Ce^{3+}	4.6	42.7	10 400	30.0	390	46
$Cs_2NaLaBr_6$:Ce^{3+}	3.9	46.9	17 000	11.3	414	61
Cs_2NaYBr_6:Ce^{3+}	3.9	44.5	9500	6.3	420	61
$Cs_2NaLuBr_6$:Ce^{3+}	4.3	52.3	5800	10.5	422	61
$Cs_2LiLaBr_6$:Ce^{3+}	3.3	44.1	60 000	2.9	410	55
Cs_2LiYBr_6:Ce^{3+}	4.1	41.5	24 000	7.0	389	85
K_2LaBr_5:Ce^{3+}	3.9	42.8	40 000	4.9	359	100
$Rb_2LiLaBr_6$:Ce^{3+}	3.9	42.1	33 000	4.8	363	26
Rb_2LiYBr_6:Ce^{3+}	3.8	35.9	23 000	4.7	385	42
$RbGd_2Br_7$:Ce^{3+}	4.8	50.6	56 000	3.8	420	43
K_2LaI_5:Ce^{3+}	4.4	52.4	57 000	4.2	401	24
GdI_3:Ce^{3+}	5.2	56.9	47 000	4.7	525	45
$Cs_3Gd_2I_9$:Ce^{3+}	4.7	57.0	2600		571	
LuI_3:Ce^{3+}	5.7	60.5	98 000	4.6	475	33
Cs_3LuI_6:Ce^{3+}	4.8	57.0	1500		429	
$Cs_3Lu_2I_9$:Ce^{3+}	4.8	57.9	22 800	9.0	556	18
Gd_2O_2S:Pr^{3+}, Ce^{3+}	7.3	61.1	40 000		511	3000
YAl_2O_3:Ce^{3+}	5.4	25.6	17 000	5.7	370	26
$Y_3Al_5O_{12}$:Ce^{3+}	4.6	35.1	17 000	3.5	550	85
$Y_3Al_5O_{12}$:Pr^{3+}	4.6	35.1	16 000		300 to 400	18
$LuAlO_3$:Ce^{3+}	8.3	64.9	11 400	23	365	17
Lu_2SiO_5:Ce^{3+}	7.4	50.2	30 000	10	420	40
Lu_2SiO_5:Pr^{3+}	7.4	50.2	2200		247	26

$Lu_2Si_2O_7:Ce^{3+}$	6.2	46.4	26 000	9.5	378	38
$Lu_{2.25}Y_{0.75}Al_5O_{12}:Pr^{3+}$	6.2	44.1	33 000	4.4		
$Lu_3Al_5O_{12}:Ce^{3+}$	6.7	44.3	12 500		510	70
$Lu_3Al_5O_{12}:Pr^{3+}$	6.7	44.3	19 000	4.6	310	20
$Gd_2SiO_5:Ce^{3+}$	6.7	45.3	12 500	7.0	430	56
$Gd_2Si_2O_7:Ce^{3+}$	5.5	41.8	40 000	6.0	372	46
$Lu_{2x}Gd_{2-2x}SiO_5:Ce^{3+}$	7.3	63	30 000 to 39 000	8	410 to 430	30 to 40

Table 3.3. Overview of some Eu^{2+} and Yb^{2+} doped scintillators. Reprinted from [37] with permission of MDPI (CC-BY SA 4.0).

Scintillator	ρ(g cm^{-3})	Z_{eff}	Light yield (photons MeV^{-1})	Energy resolution (%, at 662 keV)	λ (nm)	τ (ns)
$BaCl_2:Eu^{2+}$	3.89	49.8	19 400	8.8	402	390
$BaBr_2:Eu^{2+}$	4.78	47.8	15 700	11.0	404	585
$BaI_2:Eu^{2+}$	5.15	54.1	59 000	8.0	425	610
$BaFI:Eu^{2+}$	5.45	49.3	55 000	8.5	405	584
$BaClBr:Eu^{2+}$	4.50	44.2	52 000	3.6	407	484
$BaClI:Eu^{2+}$	4.60	49.4	54 000	9.0	410	
$BaBrI:Eu^{2+}$	5.20	50.3	97 000	3.4	412	432
$Ba_2SiO_4:Eu^{2+}$	5.47	40.9	40 000		505	582
$BaKPO_4:Eu^{2+}$	4.14	34.6	35 000		425	540
$Ba_2Si_3O_8:Eu^{2+}$	3.97	35.1	35 000		505	1296
$BaSi_2O_5:Eu^{2+}$	3.73	33.4	30 000		520	2800
$Ba_3(PO_4)_2:Eu^{2+}$	5.25	38.8	27 000		420	459
$Ba_3P_4O_{13}:Eu^{2+}$	4.10	34.9	25 000		440	669
$Ba_5Si_8O_{21}:Eu^{2+}$	3.93	34.7	20 000		453	742
$BaNaPO_4:Eu^{2+}$	4.27	34.5	20 000		450	566
$Ba_3B(PO_4)_3:Eu^{2+}$	4.17	35.4	18 000		418	698
$Ba_2ZnSi_2O_7:Eu^{2+}$	4.75	36.2	16 000		505	748
$BaAl_{10}MgO_{17}:Eu^{2+}$	3.77	24.4	16 000		459	1100
$Ba_2B_5O_9Cl:Eu^{2+}$	3.75	32.8	11 000		420	640
$Ba_2MgSi_2O_7:Eu^{2+}$	4.26	35.8	10 000		505	692
$CaF_2:Eu^{2+}$	3.40	15.5	24 000		430	940
$CaBr_2:Eu^{2+}$	3.35	35.3	36 000	9.1	448	2500
$CaI_2:Eu^{2+}$	3.96	48.0	110 000	8.0	470	790
$Cs_2BaCl_4:Eu^{2+}$	3.75	44.9	30 000		431	
$Cs_2BaBr_4:Eu^{2+}$	4.40	47.0	25 000		441	
$Cs_2BaI_4:Eu^{2+}$	4.50	54.0	17 000		462	
$CsBa_2Br_5:Eu^{2+}$	4.48	46.2	92 000		430	844
$CsBa_2I_5:Eu^{2+}$	4.90	54.0	102 000	2.6	435	1200
$CsBa_2I_5:Yb^{2+}$	4.90	54.0	54 000	5.7	414	870
$CsCaCl_3:Eu^{2+}$	3.00	44.3	18 000	8.9	445	5050

(*Continued*)

Table 3.3. (*Continued*)

Scintillator	ρ(g cm^{-3})	Z_{eff}	Light yield (photons MeV^{-1})	Energy resolution (%, at 662 keV)	λ (nm)	τ (ns)
$CsCaBr_3:Eu^{2+}$	3.72	46.6	28 000	9.3	447	6097
$CsCaI_3:Eu^{2+}$	4.06	54.0	38 500	8.0	450	1720
$CsSrCl_3:Eu^{2+}$	2.57	38.9	33 400	11.5	442	2700
$CsSrBr_3:Eu^{2+}$	3.34	42.1	31 300	9.0	448	2500
$CsSrI_3:Eu^{2+}$	3.74	51.4	65 000	5.9	450	3300
$K_2BaI_4:Eu^{2+}$	4.12	49.4	63 000	2.9	448	1500
$KBa_2I_5:Eu^{2+}$	4.47	52.2	90 000	2.4	444	1700
$SrBr_2:Eu^{2+}$	4.22	36.1	20 000	7.0	410	
$SrI_2:Eu^{2+}$	4.60	49.4	85 000	3.7	422	1200
$SrI_2:Yb^{2+}$	4.60	49.4	56 000	4.4	414	610
$SrClI:Eu^{2+}$	4.10	42.8	70 000		414	
$DSrBrI:Eu^{2+}$	4.90	44.2	47 000		418	

3.3 Semiconductor scintillators

In gamma-ray detector applications, the requirements and required physical attributes for semiconductor materials are highly strict. High atomic number to ensure sufficient gamma-ray absorption, high resistivity with high bandgap semiconductor with the low dark current without cooling, and excellent charge-transport properties to ensure full charge collection with the fewest possible charge trapping centers are all requirements for ideal semiconductors. Only a few materials, such as HgI_2, CdTe, CdZnTe, and TlBr, are available to fulfill these challenges after more than four decades of research and development; only $Cd_{0.9}Zn_{0.1}Te$ (CZT) is commercially used for detector fabrication [38–40].

3.4 Transparent ceramic scintillators

When a unique shape, size, or low cost cannot be achieved with single-crystal, transparent ceramics are frequently considered. The scintillating characteristics of SrF_2 translucent ceramics and single crystals were explored by Kato *et al* [41]. Under ^{57}Co γ-rays, the SrF_2 ceramic had a scintillation yield of 8900 photons MeV^{-1}, which was lower than a single crystal. Under 137Cs-ray irradiation, the decay time constants 4 computed for the 0.01%, 0.04%, 0.2%, and 0.5% Eu-doped CaF_2 ceramics were 1060, 714, 646, 464 ns, respectively, and light yields were 1000, 3480, 9850, 3940 photons MeV^{-1}, respectively[42]. Yanagida *et al* used the sintering approach to make a transparent optical ceramic with Pr 0.2%–1% doped $Lu_3Al_5O_{12}$ (Pr:LuAG) and compared its optical and scintillating capabilities to a single crystal equivalent created using the CZ method. For the first time, researchers discovered that the ceramic Pr 0.25% doped LuAG had a higher light yield and better energy resolution than a high-quality single crystal. For gamma-ray irradiation, the absolute light yield and energy resolution of single crystals were 180 001 000 ph

Table 3.4. A list of important organic scintillators.

Scintillator	ρ (gm cm^{-3})	Z_{eff}	Light yield (photons MeV^{-1})	λ(nm)	τ(ns)	References
Anthracene	1.25	5.24	16 000	447	30	[44]
Stilbene	1.16	5.14	8000	410	4.5	[44]
Naphthalene	0.96	5.18	2000	348	80	[44]
2,5-Diphenyloxazole	1.06	5.52	8800	405	7	[44]

Table 3.5. Summary of gamma-ray detection results based on organic-inorganic hybrid perovskites.

Compounds	Tested γ-ray source	Energy resolution	Applied bias (V)	References
$FAPbI_3$	^{241}Am	35%	23	[45]
$MAPbbr_{2.94},C_{1.06}$	^{137}Cs	12% and 6.5%	10	[46]
$MAPbI_3$	^{241}Am	12%	70	[47]
$MAPbI_3$	^{57}Co	6.8%	70	[47]

MeV^{-1} and 5.0% at 662 keV, respectively, whereas those of bulk optical ceramics were 218 001 100 ph MeV^{-1} and 4.6% at 662 keV [43].

3.5 Organic compounds and composites

Conjugated hydrocarbon compounds with an extended conjugated system of electrons in the double carbon bonds of the molecules are traditional organic scintillators. Organic scintillators have a number of advantages over inorganic scintillators, the most important of which is their low production cost. Some organic scintillators are listed in table 3.4 [44].

3.6 Organic-inorganic hybrid perovskites

For the first time, He *et al* demonstrated that solution-grown single crystals of organic–inorganic hybrid perovskite $CH_3NH_3PbI_3$ ($MAPbI_3$) may build remarkable hard radiation detectors with a high spectrum response and low dark current when used in a Schottky-type device design [45]. For the ^{57}Co 122 keV gamma-ray, the Schottky-type $MAPbI_3$ detector obtains an exceptional energy resolution of 6.8%. Due to the balanced charge collection efficiency for both electrons and holes, which is represented in the high mobility-lifetime products for both carriers, excellent detector performance is attained. $MAPbI_3$ also has extremely extended electron and hole lifetimes, as well as outstanding long-term operating stability. The Schottky-type $MAPbI_3$ detector also achieves dual-source detection of alpha-particles (5.5 MeV) and gamma-rays (59.0 keV) from the 241Am radiation source. These findings show that $MAPbI_3$ has a lot of potential as a high-performance, low-cost radiation detector material. As a result, $MAPbI_3$ is a promising option for next-generation near-room-temperature radiation detection. Recently discovered hybrid perovskites are also listed in table 3.5.

To analyze the scintillation properties of $(C_6H_5C_2H_4NH_3)_2Pb_{1-x}Sr_xBr_4$ (x = 0.1, 0.25, and 0.5) single crystals produced by the poor-solvent diffusion method, Akatsuka *et al* discovered that under ^{137}Cs gamma-ray irradiation, the pulse height spectra observed and calculated scintillation light yields were 19 700 and 18 500 ph MeV^{-1} for x = 0.1 and 0.25 samples, respectively. They also discovered good scintillation response linearity in the energy range of 22–662 keV[48].

Under gamma-ray and x-ray irradiation, Kawano *et al* examined the scintillation properties of organic–inorganic layered perovskite-type compounds containing phenethylamine. They used the poor-solvent diffusion approach to make the bulk crystal. The scintillation light production of Phe crystal was thought to be proportional to the gamma-ray energy range of 122–662 KeV [49]. Perovskites that can be processed in a solution are a new family of materials for scintillation applications. Despite the fact that research on these materials is still in its early stages. They have, nevertheless, already demonstrated extraordinary properties.

3.7 Nanostructured scintillators

In comparison to unpatterned scintillators, nanostructuring of scintillators and quantum dot (QD) or nanocrystal scintillators have been demonstrated to improve performance. Readers can find more information in reference [50].

References

[1] Yanagida T 2018 Inorganic scintillating materials and scintillating detectors *Proc. Jpn. Acad., Ser.* B **94** 75

[2] Sakai E 1987 Recent measurements on scintillator-photodetector systems *IEEE Trans. Nucl. Sci.* **34** 418

[3] Holl I, Lorenz E and Mageras G 1988 A measurement of the light yield of common inorganic scintillators *IEEE Trans. Nucl. Sci.* **35** 105

[4] Shimizu Y, Minowa M, Suganuma W and Inoue Y 2006 Dark matter search experiment with CaF_2(Eu) scintillator at Kamioka Observatory *Phys. Lett.* B **633** 195

[5] Grabmaier B C 1984 Crystal scintillators *IEEE Trans. Nucl. Sci.* **31** 372

[6] Weber M J and Monchamp R R 1973 Luminescence of $Bi_4Ge_3O_{12}$ – spectral and decay properties *J. Appl. Phys.* **44** 5495–9

[7] Melcher C L and Schweitzer J S 1992 Ceriumdoped lutetium oxyorthosilicate – a fast, efficient new scintillator *IEEE Trans. Nucl. Sci.* **39** 502–5

[8] Pidol L, Kahn-Harari A, Viana B, Virey E, Ferrand B and Dorenbos P *et al* 2004 High efficiency of lutetium silicate scintillators, Ce doped LPS, and LYSO crystals *IEEE Trans. Nucl. Sci.* **51** 1084–7

[9] Kamada K, Endo T, Tsutumi K, Yanagida T, Fujimoto Y and Fukabori A *et al* 2011 Composition engineering in cerium-doped $(Lu,Gd)_3(Ga,Al)_5O_{12}$ single-crystal scintillators *Cryst. Growth Des.* **11** 4484–90

[10] Yanagida T, Itoh T, Takahashi H, Hirakuri S, Kokubun M and Makishima K *et al* 2007 Improvement of ceramic YAG(Ce) scintillators to $(YGd)_3Al_5O_{12}(Ce)$ for gamma-ray detectors *Nucl. Instrum. Methods Phys. Res.* A **579** 23–6

[11] Yanagida T, Koshimizu M, Okada G, Kojima T, Osada J and Kawaguchi N 2016 Comparative study of nondoped and Eu-doped SrI_2 scintillator *Opt. Mater.* **61** 119–24

[12] Yanagida T, Yoshikawa A, Yokota Y, Maeo S, Kawaguchi N and Ishizu S *et al* 2009 Crystal growth, optical properties, and alpha-ray responses of Ce doped $LiCaAlF_6$ for different Ce concentration *Opt. Mater.* **32** 311–4

[13] Yanagida T, Kawaguchi N, Fujimoto Y, Fukuda K, Yokota Y and Yamazaki A *et al* 2011 Basic study of Europium doped $LiCaAlF_6$ scintillator and its capability for thermal neutron imaging application *Opt. Mater.* **33** 1243–7

[14] Yanagida T, Fujimoto Y, Miyamoto M and Sekiwa H 2014 Optical and scintillation properties of Cd doped ZnO film *Jpn. J. Appl. Phys.* **53** 02BC13

[15] Takagi K and Fukazawa T 1983 Cerium activated Gd_2SiO_5 single crystal scintillator *Appl. Phys. Lett.* **42** 43–5

[16] Yang K, Melcher C L, Koschan M A and Zhuravleva M 2011 Effect of Ca co-doping on the luminescence centers in LSO:Ce single crystals *IEEE Trans. Nucl. Sci.* **58** 1394–9

[17] Syntfeld-Kazuch A, Moszynski M, Swiderski L, Szczsniak T, Nassalski A and Melcher C L *et al* 2008 Energy resolution of calcium co-doped LSO:Ce scintillators *IEEE 2008 Conf. Record* pp 2744–50

[18] Kawamura S, Kaneko J H, Higuchi M, Yamaguchi T, Haruna J and Yagi Y *et al* 2007 Floating zone growth and scintillation characteristics of cerium-doped gadolinium pyrosilicate single crystals *IEEE Trans. Nucl. Sci.* **54** 1383–6

[19] Kawamura S, Kaneko J H, Higuchi M, Fujita F, Homma A and Haruna J *et al* 2007 Investigation of Ce-doped $Gd_2Si_2O_7$ as a scintillator material *Nucl. Instrum. Methods Phys. Res.* A **583** 356–9

[20] Suzuki A, Kurosawa S, Shishido T, Pejchal J, Yokota Y and Futami Y *et al* 2012 Fast and high-energy-resolution oxide scintillator: Cedoped $(La,Gd)_2Si_2O_7$ *Appl. Phys. Express* **5** 102601

[21] Yanagida T, Fujimoto Y, Ohgi Y, Yokota Y, Yoshikawa A and Kagi H *et al* 2010 Development and evaluations of apatite crystal scintillators *IEEE Trans. Nucl. Sci.* **57** 1308–11

[22] Igashira T, Mori M, Okada G, Kawaguchi N and Yanagida T 2017 Photoluminescence and radio luminescence properties of $(Gd_8Ca_2)(SiO_4)_6O_2$ crystals with different concentrations of Ce *J. Rare Earths* **35** 1071–6

[23] Igashira T, Mori M, Okada G, Kawaguchi N and Yanagida T 2016 Optical and radiation induced fluorescence properties of Ce: $(Gd_8X_2)(SiO_4)_6O_2$ (XFMg,Ca,Sr,Ba) crystals *Opt. Mater.* **64** 239–44

[24] Lempicki A, Randles M H, Wisniewski D, Balcerzyk M, Brecher C and Wojtowicz A J 1995 $LuAlO_3$-Ce and other aluminate scintillators *IEEE Trans. Nucl. Sci.* **42** 280–4

[25] Yanagida T, Ito T, Takahashi H, Sato M, Enoto T and Kokubun M *et al* 2007 Improvement of ceramic YAG(Ce) scintillator to $(YGd)_3Al_5O_{12}(Ce)$ for gamma-ray detectors *Nucl. Instrum. Methods Phys. Res.* A **579** 23–6

[26] Yanagida T, Kamada K, Fujimoto Y, Yagi H and Yanagitani T 2013 Comparative study of ceramic and single crystal Ce:GAGG scintillator *Opt. Mater.* **35** 2480–5

[27] Nikl M, Ogino H, Krasnikov A, Beitlerova A, Yoshikawa A and Fukuda T 2005 Photo- and radio luminescence of Pr-doped $Lu_3Al_5O_{12}$ single crystal *Phys. Status Solidi A Appl. Res.* **202** R4–6

[28] van Loef E V, Wilson C M, Cherepy N J, Hull G, Payne S A and Choong W *et al* 2009 Crystal growth and scintillation properties of strontium iodide scintillators *IEEE Trans. Nucl. Sci.* **56** 869–72

[29] Burger A, Rowe E, Groza M, Figueroa K M, Cherepy N J and Beck P R *et al* 2015 Cesium hafnium chloride: a high light yield, non-hygroscopic cubic crystal scintillator for gamma spectroscopy *Appl. Phys. Lett.* **107** 143505

[30] Saeki K, Fujimoto Y, Koshimizu M, Yanagida T and Asai K 2016 Comparative study of scintillation properties of Cs_2HfCl_6 and Cs_2ZrCl_6 *Appl. Phys. Express* **9** 042602

[31] Sakai T, Koshimizu M, Fujimoto Y, Nakauchi D, Yanagida T and Asai K 2017 Luminescence and scintillation properties of Tl- and In-doped CsCl crystals *Jpn. J. Appl. Phys.* **56** 062601

[32] Fujimoto Y, Saeki K, Yanagida T, Koshimizu M and Asai K 2017 Luminescence and scintillation properties of $TlCdCl_3$ crystal *Radiat. Meas.* **106** 151–4

[33] Saeki K, Fujimoto Y, Koshimizu M, Nakauchi D, Tanaka H and Yanagida T *et al* 2017 Luminescence and scintillation properties of Tl and Ce-doped Cs_2HfCl_6 crystals. *Jpn. J. Appl. Phys.* **56** 020307

[34] Saeki K, Wakai Y, Fujimoto Y, Koshimizu M, Yanagida T and Asai K 2016 Luminescence and scintillation properties of Rb_2HfCl_6 crystals. *Jpn. J. Appl. Phys.* **55** 110311

[35] Fujimoto Y, Saeki K, Yahaba T, Tanaka H, Yanagida T and Koshimizu M *et al* 2016 Photoluminescence and radiation response properties of Ce_3D-doped $CsCaCl_3$ crystalline scintillator *Phys. Scr.* **91** 094002

[36] Fujimoto Y, Koshimizu M, Yanagida T, Okada G, Saeki K and Asai K 2016 Thallium magnesium chloride: a high light yield, large effective atomic number, non-hygroscopic crystalline scintillator for X-ray and gamma-ray detection *Jpn. J. Appl. Phys.* **55** 09030

[37] Maddalena F *et al* 2019 Inorganic, organic, and perovskite halides with nanotechnology for high-lighted yield X- and gamma-ray scintillators *Crystal* **9** 88

[38] Owens A 2012 *Compounds Semiconductor Radiation Detectors* (Boca Raton, FL: CRC Press)

[39] Eisen Y, Shor A and Mardor I 1999 CdTe and CDZnTe gamma ray detectors for medical and imaging systems *Nucl. Instrum. Methods Phys. Res. Sect.* **428** 158–70

[40] Roy U N, Giuseppe S and Camarda S 2019 Evaluation of CdZnTeSe as a high-quality gamma-ray spectroscopic material with better compositional homogeneity and reduced defects *Sci. Rep.* **9** 7303

[41] Kato T, Kawano N, Okada G and Kawaguchi N 2018 Scintillating properties of SrF_2 translucent ceramics and single crystal *Optik* **168** 956

[42] Lan Y, Mei B, Li W, Xiong F and Song J 2019 Preparation and scintillation properties of Eu2+:CaF2 scintillation ceramics *J. Lum.* **208** 183

[43] Yanagida T, Fujimoto Y, Kamada K and Totsuka D *et al* 2012 Scintillation properties of transparent ceramic Pr:LuAG for different Pr concentration *IEEE Trans. Nuc. Sci.* **58** 2146

[44] Maddalena F, Tajahana L and Xie A *et al* 2019 Inorganic, organic, and perovskite halides with nanotechnology for high-light yield X- and gamma-ray scintillators *Crystals* **9** 88

[45] Wei H and Huang J 2019 Halide lead perovskites for ioning radiation detection *Nat. Commun.* **10** 1–12

[46] Wei H and DeSantis D *et al* 2017 Dopant compensation in alloyed $CH_3NH_3PbBr_{3-x}Cl_x$ perovskite single crystals for gamma-ray spectroscopy *Nat. Mater.* **16** 826

[47] Yakunin S, Dirin D N and Shynkarenko Y *et al* 2016 Detection of gamma photons using solution-grown single crystals of hybrid lead halide perovskites *Nat. Phot.* **10** 585

[48] Akatsuka M and Kawano N *et al* 2020 Development of scintillating 2D quantum confinement materials – $(C_6H_5C_2H_4NH_3)_2Pb_{1-x}$ Sr_xBr_4 *Nuc. Inst. Meth. Phy. Res.* A **954** 161372

[49] Kawano N, Koshimizu M and Okada G *et al* 2017 Scintillating organic-inorganic layered perovskite-type compounds and the gamma-ray detection capabilities *Sci. Rep.* **7** 14754

[50] Dujardin C and Auffray E *et al* 2018 Needs, trends, and advances in inorganic scintillators *IEEE Trans. Nucl. Sci.* **65** 1977–97

Chapter 4

Material processing for radiation detectors

4.1 Introduction

The growth and fabrication of critical materials for use in *scintillating radiation detectors* in a variety of forms are presented, including bulk single crystals, polycrystalline ceramics, thin films, thick films, and composites. The fundamentals of crystal growth and materials processing techniques are described with the methodology's related illustration. However, based on the characteristic of a starting material and on applicability and requirements, the processes's parameters are selected. However, single crystals are sill the dominant bulk form of inorganic scintillator materials. Due to the limitations imposed by crystal growers, researchers have been looking for and developing alternate media, such as thin- and thick-films, composites, transparent ceramics, and nanomaterials. Table 4.1 describes some of the critical scintillating materials, their form, and methods utilized to fabricate them.

Crystal growth is an art and science of growing crystals that are pillars of modern technological developments, especially in the electronic device industry [1]. Crystal growth technology acts as a bridge between science and technology. The scintillating sensors, semiconducting sensors for radiation detection, lasers, semiconducting devices, computers, magnetic and optical devices warrant good quality crystals in bulk or film or noncrystalline forms, including pharmaceuticals and a host of other devices. Crystal-growth requires technical skills in chemistry, electronics, physics, and materials science. The formation of a solid crystalline phase can only occur if some degree of supersaturation or super-cooling has been achieved first in the crystal growth system. The crystallization process contains three necessary steps:

- attainment of supersaturation or super-cooling;
- formation of critical-sized nuclei, i.e., nucleation;
- growth of the nuclei into a crystal.

After the system viz. vapor, melt, or solution is in thermal equilibrium with its nucleus, the latter is capable of growing by attaching more and more material to it very

doi:10.1088/978-0-7503-2508-0ch4

Table 4.1. Summary of essential scintillators, forms, and fabrications techniques [1–10].

S. No	Crystal name	Form	Growth method utilized
1	Alkali-halides (Li, Na, K, Rb, and Cs–F, Cl, Br, and I)	Single crystal	Bridgman–Stockbarger (BS)
2	PbI_2, HgI_2, TlBr	Single crystal or polycrystalline films	BS, Czochralski (CZ)and vapor-phase-grown or vacuum thermal evaporation or screen printing
3	Lanthanide halides, cerium-activated lanthanum chloride; $LaCl_3$:Ce; $LaBr_3$	Single crystal	BS
4	$RbGd_2Br_7$:Ce	Single crystal	BS
5	LuI_3:Ce	Single crystal	BS
6	Lu_2SiO_3 (LSO); LSO:Ce^{3+}	Single crystal	CZ method with iridium crucible
7	LSO	Powder/film	Sol–gel via tetraethyl orthosilicate as the precursor
8	Complex oxides: gadolinium silicate (GSO); bismuth germinate (BGO); cadmium tungstate (CWO); lead tungstate (PWO)	Single crystal/film	Czochralski (CZ) technique/screen printing
9	Cerium activated lutetium gadolinium silicon oxide ($Lu_{0.4}$ $Gd_{1.6}$ SiO_5;Ce)	Single crystal	CZ technique
10	Cadmium telluride (CdTe); cadmium zinc telluride (CZT)	Single crystal	Traveling heater method (THM); Bridgman; high-pressure Bridgman (HPB)
11	Silicon, germanium, and diamond	Single crystal or deposition	Chemical vapor deposition (CVD); CZ method
12	Gallium arsenide (GaAs)	Epitaxy layer	Liquid-phase epitaxy (LPE)

slowly via a process called 'growth.' Nucleation may occur spontaneously (homogeneously), or it may be induced artificially (heterogeneously). The growth of crystals ranges from a small, inexpensive technique to a complex, sophisticated, and expensive processes, and crystallization time varies from minutes to hours, days, or months. It is pertinent to state that the transport of crystal constituents in the solid, liquid, or vapor phase creates single crystals. Thus, the classification of crystal growth techniques is solid-growth, vapor-growth, melt-growth, and solution-growth, accordingly.

4.2 Czochralski (CZ) technique

The Czochralski (CZ) technique, often known as 'crystal pulling,' was created to determine the rate of metal crystallization. It entails melting a starting powder in a crucible consisting of platinum, iridium, graphite, or ceramic, depending on the specifications. With the melt at an acceptable temperature, a revolving rod with a tiny seed-crystal attached on one end is dropped into the crucible until it barely touches the melt surface, then slowly withdrawn (1–2 mm h^{-1}). At the contact between the melt and the seed, crystallization occurs in two ways:

- Surface tension draws a small amount of the melt from the crucible and onto the seed.
- Heat conduction permits the solid on the seed crystal to outspread slightly into the melt, ensuring that enough material is drawn out to grow the developing crystal ever larger.

The crystal growth will continue until the crucible's entire contents have been removed and fed to the rod. The draw rate is usually between 1 mm h^{-1} and 10 cm h^{-1}. The diameter and height of CZ crystals can be huge. Pulling is used to create a range of technologically important crystals, such as pure silicon, ruby, sapphire, YAG, GGG, and a variety of rare oxides. The CZ procedure is depicted in figure 4.1.

1. In a crucible, a contained charge is heated to slightly above its melting point. Above the crucible is a pull-rod with a chuck carrying a seed crystal at its bottom end. The seed crystal on the rod is dipped into the molten metal. The temperature is then regulated until the meniscus is supported by a seed crystal. The pull rod is then steadily raised at an optimum rate, allowing the seed crystal to grow without detaching. Nucleation is usually started on a twisted platinum wire when a seed is not available. The seed serves as both a heat sink and a nucleation site, allowing the latent heat of solidification to

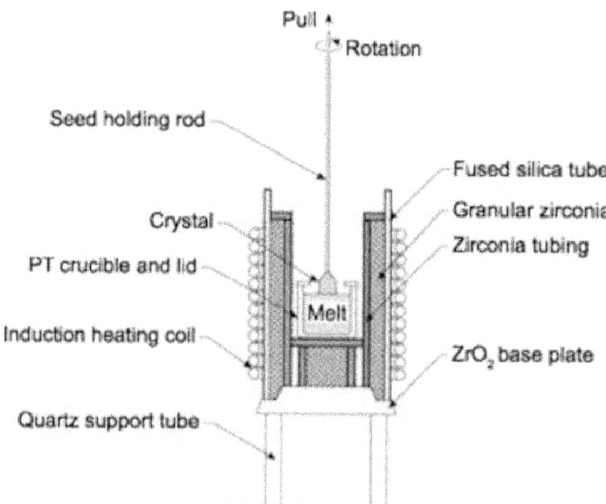

Figure 4.1. A schematic diagram of a CZ crystal growth setup. Reproduced from [1] with permission of SPIE.

escape (the solidified fraction at the surface of the seed will reproduce its single-crystal structure). To ensure symmetrical heating of the melt, the seed crystal is frequently turned in the counter-clockwise direction, and the crucible is occasionally rotated as well. Cylindrical growing crystals are common under these conditions. The following features and components are required for a versatile crystal-pulling apparatus: A rigid, vibration-free mainframe.

2. A seed crystal holder/chuck and pull head and control of its rotation rates.
3. A range of crucibles to hold the melt that is compatible with the surrounding atmosphere.
4. Insulating materials that act as a thermal barrier around the crucible and control the thermal field in the crystallization area.
5. A heating source and temperature-controlled system.
6. An arrangement to control the crystal diameter.
7. A CCD camera or other optical system to view the crystal.

4.2.1 Parameters, advantages, and disadvantages

Because the pull rate is the velocity at which the seed crystal is dragged up, it is a precisely controlled variable.

Rotation rate: Rotation rates typically range from 5 to 40 rpm, depending on the crystal diameter and desired configuration.

Thermal gradients: For the crystallization process to work, there must be a non-uniform temperature distribution inside and above the crucible: the temperature in the liquid is obviously higher than the melting point T_m, and the temperature in the crystal is lower, with the interface at T_m.

4.2.1.1 Advantages

- Growth can be performed with a free surface (volume changes are accommodated);
- The crystal can be viewed;
- Forced convection is easy to impose;
- High throughput; huge crystals are possible;
- It is possible to obtain high crystalline perfection;
- Excellent radial uniformity.

4.2.1.2 Disadvantages

- High vapor pressure materials cannot be developed;
- Batch procedures are difficult to adapt for continuous growth, resulting in axial segregation;
- The crystal must be rotated, and crucible rotation is desirable;
- The procedure necessitates constant monitoring (for seeding and necking) as well as sophisticated management.

4.2.2 Si crystal formation procedure recipe

- The crucible is filled with polycrystalline silicon, and the furnace is heated above the melting point of silicon (1420 °C).
- A <111>-oriented seed crystal is suspended over the crucible in the seed holder. The seed is inserted into the melt. Part of it melts, but the tip of the remaining seed crystal still touches the liquid surface. It is slowly withdrawn.
- After a short duration (1–3 min), pulling begins at a slow rate. The new crystal will grow with a smaller diameter (slightly less than the seed diameter).
- The pulling rate (mm h^{-1}) and rotation rate are increased to the final value, the diameter of the crystal will decrease, producing a narrow neck. The length of the neck is 5–10 times the diameter of the seed.
- After the neck is completed, the melt temperature is slowly lowered, and the diameter of the crystal will then increase. Then the melt temperature is kept constant, and the crystal attains its final diameter.
- Growth at a constant diameter continues until the desired length is achieved. After that, the growth is terminated by either pulling the crystal sharply or increasing the melt temperature.
- After growth is finished, the system is cooled down. The grown crystal is taken out of the system.

4.2.3 Liquid-encapsulated Czochralski method

The liquid-encapsulated Czochralski (LEC) method is identical to the CZ method, except that the crystals are produced under pressure and the melt surface is coated with boron oxide (B_2O_3), as illustrated in figure 4.2. The starting ingredients (pre-synthesized polycrystalline chunks or, in the case of semi-insulating GaAs, elemental Ga and As) are combined with a B_2O_3 pellet in the growing crucible. The crucible is heated and placed within a high-pressure crystal puller. The B_2O_3 melts at 460 °C and condenses into a thick, viscous liquid that coats the whole melt, including the

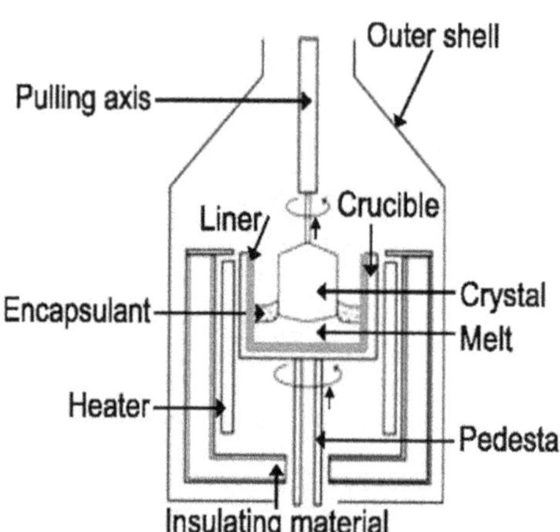

Figure 4.2. Schematic of a liquid encapsulated CZ furnace. Reproduced from [1] with permission of SPIE.

crucible. Sublimation of the volatile group-V element is prevented by this layer and the pressure in the crystal puller.

The temperature is gradually raised until the chemical begins to synthesize (specifics vary depending on the material being produced). A seed crystal is dipped into the melt through the boron-trioxide layer. A single crystal propagates from the seed after it is spun and gently extracted. CCTV is used to monitor growth, and weights, temperatures, and pressures are measured at regular intervals.

4.2.3.1 Advantages

- Materials with high vapor pressure can be grown;

- Retains most CZ advantages;
- B_2O_3 melt does not react with the crucible or ambient, but it does dissolve oxides (e.g., Ga_2O_3).

4.2.3.2 Disadvantages

- Some loss of volatile constituent;
- Contamination by B_2O_3 B_2O_3 is too viscous below 1000 °C;
- The encapsulant is opaque near the end of growth.

4.3 Bridgman–Stockbarger technique

With the Bridgman–Stockbarger (BS) process, materials that melt congruently, do not break down before melting and do not undergo phase transformation between the melting point and room temperature can be formed as a single crystal (figure 4.3).

The to-be-grown material is encased in a glass or quartz tube and suspended in a furnace with an appropriate growth gradient. The conical tip of the ampoule aids in

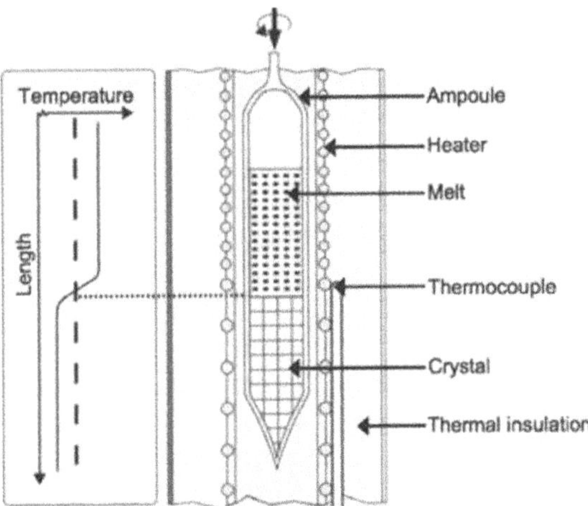

Figure 4.3. Schematic of a Bridgman–Stockbarger technique with temperature profile. Reproduced from [1] with permission of SPIE.

the nucleation of a single crystal. The growth ampoule is progressively moved from the hot zone to the cold zone after the substance has melted. The melt is filled into the lower half of the ampoule via a capillary tip; as it is lowered into the gradient, the seed germinates and grows throughout the entire melt zone of the ampoule. The pace at which the ampoule is decreased is determined by the material, i.e., the molecular and crystalline complexity.

4.3.1 Advantages and disadvantages of the BS technique

4.3.1.1 Advantages

- In confined growth, the shape of the crystal is defined by the container;
- Radial temperature gradients are not needed to control the crystal shape;
- Low thermal stresses result in a low level of stress-induced dislocations;
- Crystals may be grown in sealed ampules (the stoichiometry of melts with volatile constitutes is easy to control);
- Relatively low level of natural convection; the melt is exposed to stabilizing temperature gradients;
- Requires little attention (maintenance).

4.3.1.2 Disadvantages

- Confined growth; the container exerts pressure on the crystal during cooling;
- Hard to observe the seeding process and the growing crystal;
- Level of natural convection changes as the melt is depleted; forced convection is hard to impose;
- Ampule and seed preparation, sealing, etc do not tend toward high throughput production.

4.3.2 Recipe for bulk lead-iodide growth

Due to its high density (6.2 g cm^{-3}) and broad bandgap (2.3–2.5 eV), lead iodide (PbI_2) has been utilized for x-ray and gamma-ray detection since the 1970s, allowing detectors to function at, or even above, ambient temperature. In terms of electrical properties and anisotropic (layered) structure, PbI_2 is similar to mercury iodide (HgI_2). However, the latter undergoes destructive phase transformation at 130 °C, whereas the former is stable up to its melting point (410 °C), has a lower vapor pressure, and can be grown directly from the melt. The following are the steps that lead to growth:

1. At $T > T_m$, the polycrystalline ingot is melted in the hot zone, or zone 1.
2. For a period of t_h, the melt is homogenized.
3. Crystallization starts with lowering the ampoule at a rate of d (mm h^{-1}); it goes downward (relative to the furnace) across a distance of zB.
4. At $T < T_m$, the ampoule is lowered to the cold region's end, or zone 2.
5. The crystal is cooled while remaining in zone 2 in a fixed position. 2–3 cooling steps at a cooling rate R_c are done, each followed by a holding period t_c, depending on the experiment.

6. The furnace is turned off, and the crystal naturally cools to room temperature (20 °C) (table 4.2).

A transparent furnace allows one to observe (a) nucleation (if multiple nucleation sites occur, solidification can be restarted); (b) the melt–solid interface; (c) convection (via an index of refraction changes with temperature); (d) internal temperature (tomographically with a thermal imager); and (e) crystal defects, depending on the optical properties.

4.3.2.1 Lead-iodide growth setup
The two-zone transparent furnace consists of an inner quartz muffle tube surrounded by two resistive heating elements whose leads are connected externally to two Eurotherm® 2404 temperature controllers with a high-limit unit for precise control of the furnace. A gold-coated Pyrex tube surrounds the furnace to monitor the growth at higher temperatures. A DC motor lowers the quartz ampoule. The effective traveling length along the vertical axis z of the furnace is 608 mm. Hardened ceramic rings, with ceramic fiber cloth wrapped around the top and bottom of the furnace, are used to obtain a better thermal profile. During the growth process, the lower furnace opening is closed by a piece of ceramic fiber to prevent a chimney effect. Temperature control of the zones is achieved by four k-type thermocouples (T_1–T_4) placed within the windings of the resistance furnace (figure 4.4).

4.3.3 High-pressure vertical Bridgman method (HPB)

For compounds such as CdTe and CdZnTe with high melting points and different vapor pressure, a high-pressure vertical Bridgman method (HPB) is usually utilized. The HPB system is a variant of the conventional BS furnace kept inside a high-pressure chamber.

4.4 Hydrothermal crystal growth

Calcite and quartz are insoluble in water, although they are soluble under high temperatures and pressures. The hydrothermal method is the name given to this type of crystal formation. Temperatures typically range from 400 °C to 600 °C, with high pressure involved (hundreds or thousands of atmospheres). Steel autoclaves with gold or silver linings are commonly used for growth (figure 4.5). Autoclaves are classified as low-, medium-, or high-pressure devices based on the amount of pressure they produce. A temperature differential between the nutrient and growth zones provides the required concentration gradient for growth. The demand for high

Table 4.2. The growth parameters utilized.

T_{Z1}(°C)	T_{Z2}(°C)	ΔT(°C)	t_R (h)	Δt_h (h)	ΔT_A (°C)	Δt_c (°C h^{-1})	v_d (mm h^{-1})
520	470	50	6.5	24	390	10	2

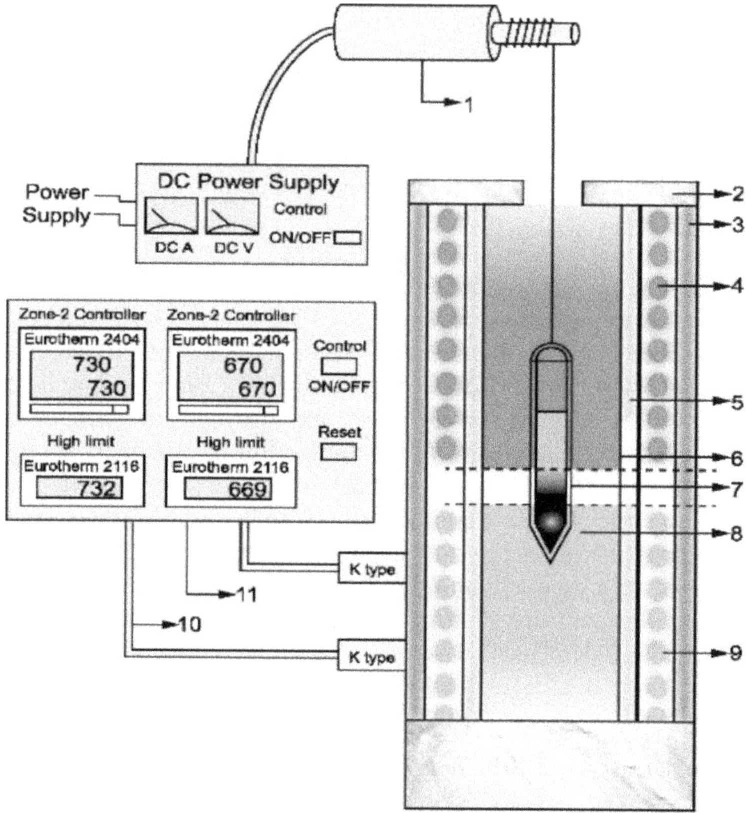

Figure 4.4. Schematic of a complete BS system: (1) DC motor-lowering unit; (2) ceramic flange end plates; (3) transparent tube; (4) hot zone ($>T_m$); (5) quartz tube; (6) molten PbI_2; (7) quartz ampoule; (8) PbI_2 crystal; (9) cold zone ($<T_m$); (10) thermocouple (k-type); (11) two-zone temperature controller. Reproduced from [1] with permission of SPIE.

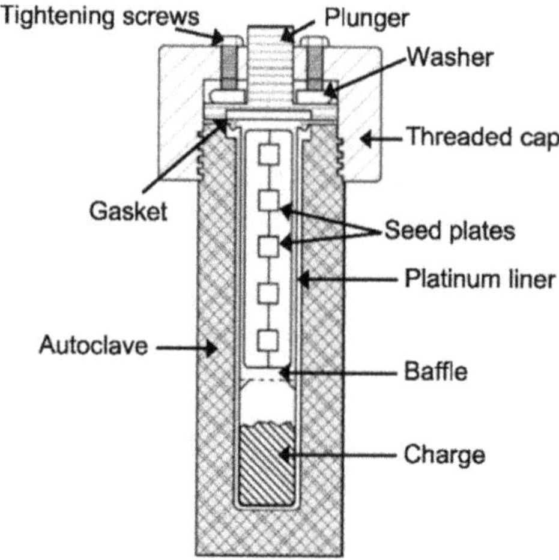

Figure 4.5. A schematic of a hydrothermal autoclave. Reproduced from [1] with permission of SPIE.

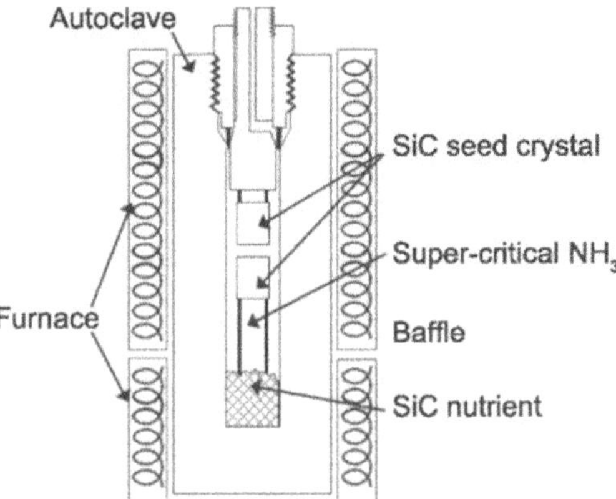

Figure 4.6. A schematic of an ammonothermal crystal growth technique. Reproduced from [1] with permission of SPIE.

pressure poses practical challenges, as this process only yields a few high-quality crystals with huge size. Quartz is a great example of hydrothermal crystallization in industry. The frequent inclusion of OH ions into the crystal, however, is a severe disadvantage of this approach, rendering it unsuitable for many applications.

4.4.1 Ammonothermal crystal growth

The ammonothermal method is a straight forward counterpart of hydrothermal technology that is used in α-quartz crystal production (figure 4.6). GaN-containing feedstock is dissolved in supercritical ammonia solution in one zone of the high-pressure autoclave, transported through the solution via convection, and finally crystallized in the second zone on GaN seeds. Typical pressure and temperatures involved are 0.1–0.3 GPa and 500 °C–900 °C, respectively. The crucial step is the choice of **mineralizer**, i.e., the ionic substance added to the reaction zone to increase the reversible dissolution of GaN in ammonia.

4.4.2 Advantage and disadvantages of a hydrothermal technique

4.4.2.1 Advantages

- Can grow large-diameter seeds with excellent structural properties;
- The repeatable and highly controlled re-crystallization process at close-to-equilibrium conditions either by dissolution or crystallization of seeds by a simple change of thermal conditions;
- Relatively low growth temperatures;
- Excellent scalability of the process with the size of autoclaves.

4.4.2.2 Disadvantages

- Dissolution of nutrients into supercritical ammonia;
- Transport of GaN onto the seed crysta;l
- Precipitation of GaN on seed(s).

4.5 Vertical floating zone melting

A heater and a limited region on the crystal gently translate a polycrystalline crystal ingot. In the floating zone procedure, the crystal will become molten (floating zone) (figure 4.7). At the liquid–solid interface, contaminants permeate from the solid zone into the liquid region and separate at the ingot's end. At the start of the procedure, a seed crystal with a certain orientation must be translated into touch with the molten region. This phase can be done with a variety of heating methods, such as an induction coil, a resistance heater, or, more recently, an optical heating system with high-power halogen lights and ellipsoidal mirrors. Simultaneously, silicon (Si) is refined and a crystal is produced.

4.5.1 Horizontal zone melting

In a non-conservative process, the material is added to the molten region. Only a small part of the charge is molten (except the seed). An axial temperature gradient is imposed along with the crucible, and the molten zone (interface) is advanced by moving the charge or the gradient (figure 4.8).

4.5.2 Advantages and disadvantages of zone melting

4.5.2.1 Advantages

- The charge is purified by the repeated passage of the zone (zone refining).
- Crystals may be grown in sealed ampules or without containers (floating zone), which results in high purity of the grown crystal.
- It can also be applied for growing single crystals of various congruently and incongruently melting materials (oxides).
- Steady-state growth is possible.
- Zone leveling is possible; can lead to superior axial homogeneity.
- The process requires little attention (maintenance).
- Simple: no need to control the shape of the crystal.
- Radial temperature gradients are high.

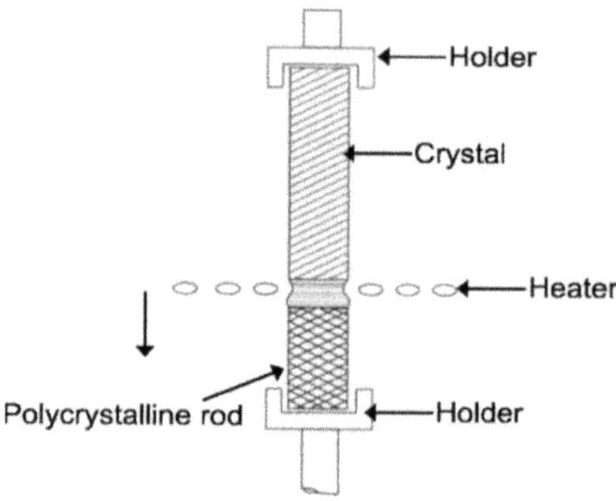

Figure 4.7. A schematic of a vertical floating zone melting technique. Reproduced from [1] with permission of SPIE.

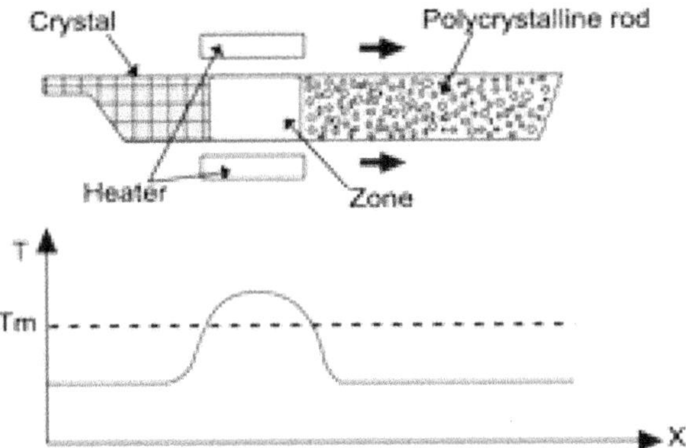

Figure 4.8. A schematic of a horizontal zone melting technique with temperature profile. Reproduced from [1] with permission of SPIE.

4.5.2.2 Disadvantages

- Confined growth (except in floating zone).
- Hard to observe the seeding process and the growing crystal.
- Forced convection is hard to impose (except in floating zone).
- In the floating zone, materials with high vapor pressure cannot be grown.

4.6 The vapor phase technology

When the molecules of a gas attach themselves to a surface and form a crystal arrangement, crystals can be formed. The average number of molecules in the gas and solid states is constant at constant temperature and equilibrium conditions: molecules leave the gas and attach to the surface at the same rate that they leave the surface to become gas molecules. The gas–solid chemical system must be in a non-equilibrium state, with too many gaseous molecules for the given pressure and temperature parameters, for crystals to develop. This is known as supersaturation, which means that molecules are more likely to escape the gas and deposit on the container's surface. Supersaturations can be created by keeping the crystal at a lower temperature than the gas. Seeding is a crucial stage in crystal growth, in which a small piece of crystal with the correct structure and orientation (the seed) is placed into the growing container. The gas molecules prefer the surface to the walls and deposit there first. When the molecule reaches the seed's surface, it moves across it in search of a suitable attachment point. One molecule and one layer at a time, this growth occurs. Because the crystals are produced at temperatures considerably below the melting point to limit the density of flaws, the procedure is lengthy; growing a small crystal takes days.

4.6.1 Advantage and disadvantage of vapor phase techniques

4.6.1.1 Advantage

- Crystals of high purity can be grown.

4.6.1.2 Disadvantage

- Vapor growth is a slow process.

4.6.2 Tube-based vapor systems

Vapor growth can be performed in two ways:

In a *close-tube system*, the material on sublimation is transported directly; the driving force is the temperature gradient (figure 4.9). The source material is kept at one end of the sealed tube (generally quartz) and heated to a temperature so that material sublimes. The vapor is transported to the other end of the tube, which is at a lower temperature, where it condenses to form crystallites.

In an *open-tube system*, a carrier gas is additionally used to transport the growth species from the source region to the growth region (figure 4.10). The source material is vaporized at the hot end of the tube, and an inert gas flows through the tube, causing transport of vapor molecules from the higher temperature to lower temperature side where they deposit/crystallize.

The tube can be dismounted, so it is easier to remove the grown crystals. Charge material can be introduced into the system just before the start of the growth run. This arrangement also provides control over doping at different stages and the transport rate.

4.6.3 Recipe for ZnSe crystal

Crystals of zinc selenide (ZnSe) can be created from the vapor phase as follows:

- A ZnSe seed single crystal and ZnSe polycrystalline source are sealed in a quartz ampoule.
- The crystal is grown in a low-pressure, inert gas at ~1100 °C.
- The seed crystal is settled in a lower-temperature position, and the polycrystalline source material is settled in a higher-temperature region.

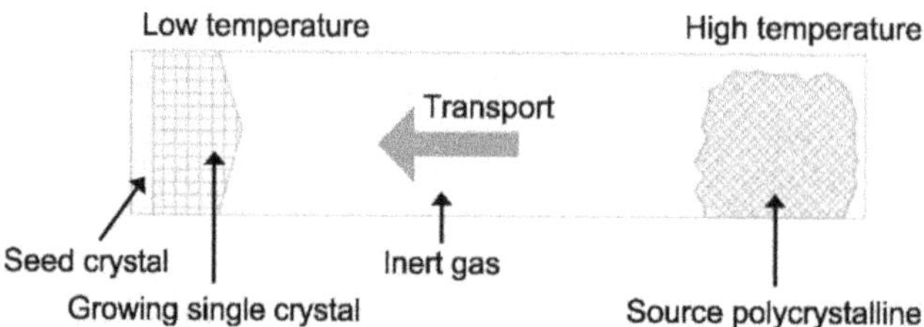

Figure 4.9. A schematic of a close-tube-based crystal growth system. Reproduced from [1] with permission of SPIE.

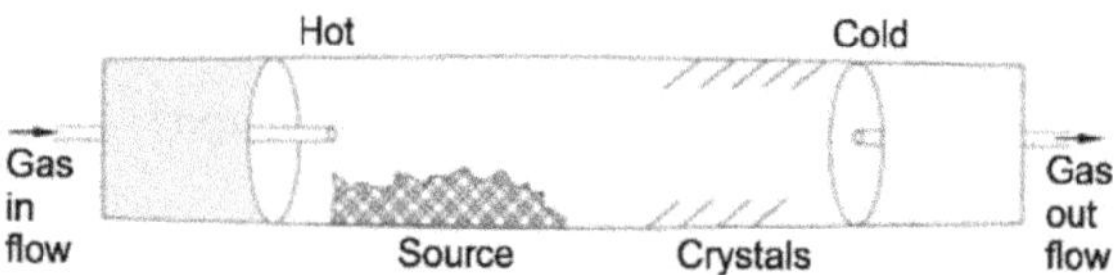

Figure 4.10. A schematic of an open-tube vapor growth setup. Reproduced from [1] with permission of SPIE.

- The source gas sublimated from the solid source material is transported to the seed crystal according to the temperature dependence of the equilibrium constant.

The crystal is grown on the seed crystal according to the following equation:

$$2ZnSe(solid) \rightarrow 2Zn(gas) + Se_2(gas)$$

The process uses an equilibrium state so that the ZnSe crystal can be easily obtained.

4.6.4 Recipe for SiC crystal growth

Silicon carbide (SiC) is a unique semiconductor material for devices in high-power, high-frequency, high-temperature, and intense-radiation applications due to its exceptional mix of physical and electrical features..

The SiC seed is fastened to the top lid of the crucible, and the source material (SiC) is held at the bottom of the crucible (figure 4.11). The distance between the seed and the source is usually around 20 mm. The crucible is heated by induction. The system's operating temperatures vary from 1800 °C to 2600 °C. A linear temperature difference between the source and the seed causes Si and C-containing species to be transported from the source to the seed in the vapor phase. Temperature gradients of 1.0 °C–2.5 °C per mm are common. The temperature at the bottom of the crystal is usually approximately 2300 °C, whereas the temperature at the seed is usually around 2100 °C. The supersaturation of a vapor phase species near the seed surface aids the crystallization process.

4.7 Preparation of ceramics

Ceramic oxide compositions are frequently created by mixing all of the constituents as oxides or compounds, such as carbonates and nitrates (which quickly decompose to oxides), and then calcinating the mixture at a temperature that allows for considerable cation inter-diffusion. After that, the calcine is carefully crushed and compressed into the desired shape before being sintered at a temperature a few hundred degrees above the calcination temperature. [10]. To calcine PZT, for example, the beginning raw ingredients PbO, TiO_2, and ZrO_2 are combined in a

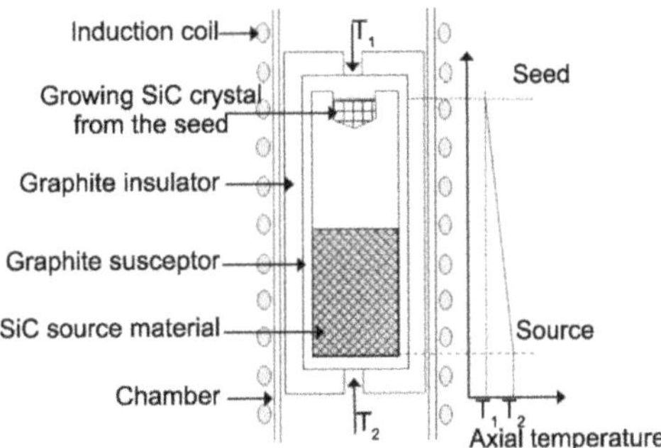

Figure 4.11. A schematic of a setup for SiC crystal growth. Reproduced from [1] with permission of SPIE.

2:1:1 molar ratio, crushed, and then calcined at 800 °C in ambient air to get the perovskite phase. The calcining temperature is important because it affects the density of the final product, as well as its electromechanical and other properties. However, calcining PZT at T > 800 °C may result in the loss of lead, which will have a negative impact on the electrical characteristics. To produce the finest electrical and mechanical properties, calcination at the proper temperature is required. The lumps are milled after they have been calcined. Before they can be sintered, the green bodies must have a particular minimum density. Various processes, such as powder compaction, slip-casting, and extrusion, can be used to achieve the appropriate shape and minimal green density. The type of powder utilized, particle size distribution, agglomeration state, intended shape, and component thickness are all factors to consider. The green bodies are gradually heated to between 500 °C and 600 °C after shaping to remove any leftover binder. To allow the gases to leave slowly without developing fractures and blisters in the ceramic sections, the binder burnout rate should be 1 °C–2 °C min^{-1}. The samples are heated to a higher temperature after the binder burnout to allow sintering to occur. For effective densification without aberrant grain formation, the sintering temperature and time should be optimal. Oxide ceramics must be sintered in an oxidizing environment or in the open air. Lead loss occurs at temperatures exceeding 800 °C for lead-containing ceramics such as PZT, PT, PLZT, and others. The samples are held in a sealed crucible with saturated vapor in it to reduce lead loss during sintering. The manufacturing of ceramics and sample preparation for characterization are depicted in figure 4.12.

4.7.1 Recipe for Pr: LuAG transparent ceramic

Aluminum, lutetium, and praseodymium water solutions are combined together to make the scintillator Pr:LuAG. The resulting aqueous solution is combined with an aqueous solution of ammonium hydrogen carbonate drop by drop. The stages of filtering and water washing are then repeated multiple times, and the resulting powder is dried at 120 °C for two days. To make raw LuAG oxide powder, the obtained precursor of lutetium, aluminum hydrate, and carbonate is calcined at around 1200 °C. The powder is then combined for two days in a ball mill with solvent, binder, and dispersion medium. The milled slurry is then poured into a

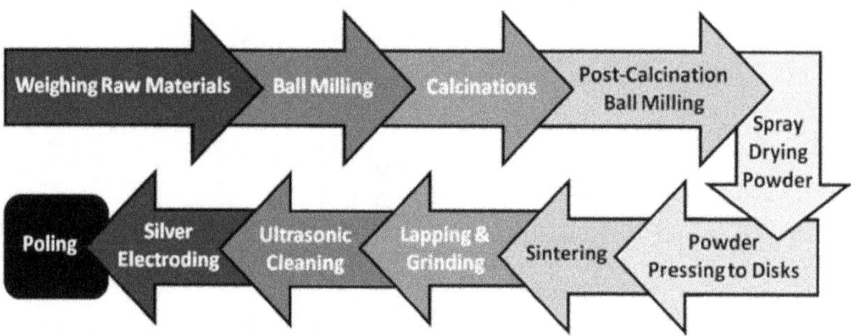

Figure 4.12. Flowchart of the fabrication and processing of ferroelectric ceramics.

gypsum mold and cured to achieve the required shape. High density and homogenous green bodies can be achieved using the slip casting forming method. The material is sintered under vacuum at 1700 °C for 20 h after the organic components have been removed by heating/calcinations. Pr:LuAG ceramic samples are very transparent after annealing.

4.8 Thin and thick-film preparation techniques

For a long time, thin-film manufacturing has piqued people's interest. Since then, several deposition methods have been developed. The related techniques can be classified into two classes: solution deposition and non-solution deposition.

4.8.1 Non-solution-deposition techniques

4.8.1.1 Sputtering

Because of its convenience and the high melting point of PZT materials, sputtering methods (DC, RF magnetron, ECR magnetron, and ion beam) have been the most extensively used approaches for fabricating ferroelectric thin films. The material to be sputtered is established as a sputtering target, which serves as the cathode in the electric circuit, during the sputtering process. The substrates to be coated are put on a grounded anode that is set at a specific distance from the target. The target could be a ceramic target with the needed compositions, or a succession of individual pieces positioned on a revolving platform. During the sputtering process, compositional ions are separated from the target's surface and deposited on the substrates.

In the preparation of PZT films, both cold and hot stages are used. Because each element's sputter rate is variable, stoichiometric control is a major issue with the sputtering process. Figure 4.13 depicts the basic concept of a sputtering principle.

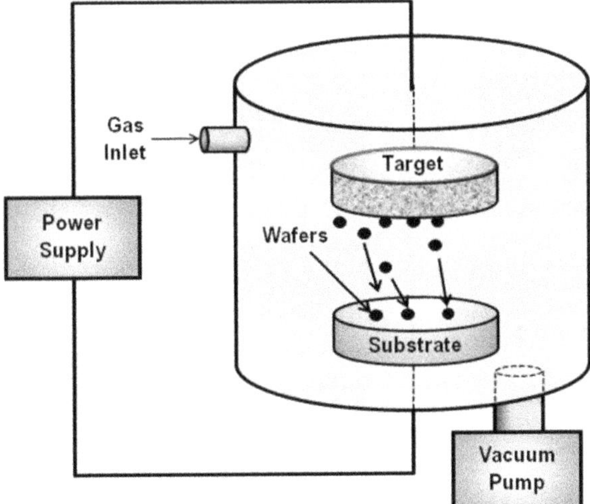

Figure 4.13. Diagram of a simple sputtering system.

4.8.1.2 Laser ablation

Pulse laser ablation deposition (PLAD) is a promising method for creating ferroelectric films due to its high deposit rate and stoichiometric control. This method can be used to generate stoichiometric films even with a five-element target material. According to research, the compositional fluence homogeneity and the target-substrate distance are the two most important characteristics that affect the quality of PZT thin films. The higher the fluence, the faster the deposition rate, but the lower the film quality, the larger the particles emitted by the target. Particle-free PZT films have yet to be produced using the PLAD method. A simplified system configuration for generating thin films with PLAD is shown in figure 4.14.

4.8.1.3 Chemical vapor deposition

Chemical vapor deposition (CVD) is the process of transporting volatile compounds to the reaction chamber, where they are exposed to the substrates, using a carrier gas. Chemical reactions occur on or near the surface of the substrates, resulting in a solid phase that condenses on the substrates. The usage of energy sources can considerably improve the CVD process. such as a laser beam, electron beam, or ion beam. The plasma, formed by a focused energy source, accelerates the decomposition of gases and improves the quality of the films. Called plasma-enhanced CVD (PECVD), this technique can provide good-quality PZT films at low processing temperatures. A schematic of a CVD apparatus is shown in figure 4.15.

Metal–organic compounds are used as volatile reactants in the metal–organic CVD (MOCVD) process. Because of its utility, high potential, and recent development of new source materials, interest in adopting this technology to produce PZT films has recently surged. MOCVD has several advantages, including composition and crystalline structural control, a fast growth rate, and excellent step coverage. On a 6–8-inch silicon wafer, uniform PZT and PLZT thin films may be generated with a film thickness variation of less than 1.5% and Pb/(Zr + Ti) and

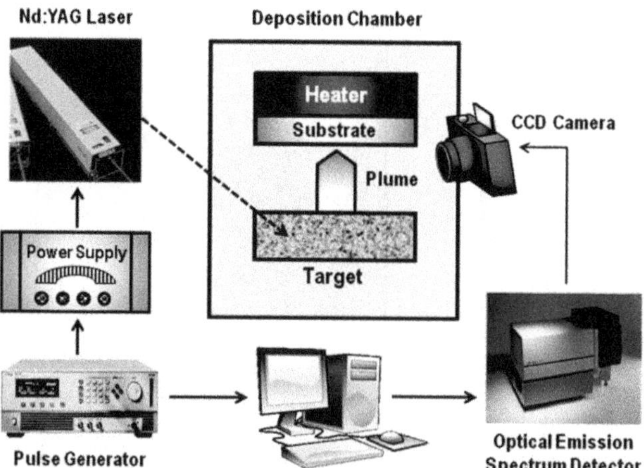

Figure 4.14. Experimental setup for pulsed laser ablation deposition (PLAD) technique.

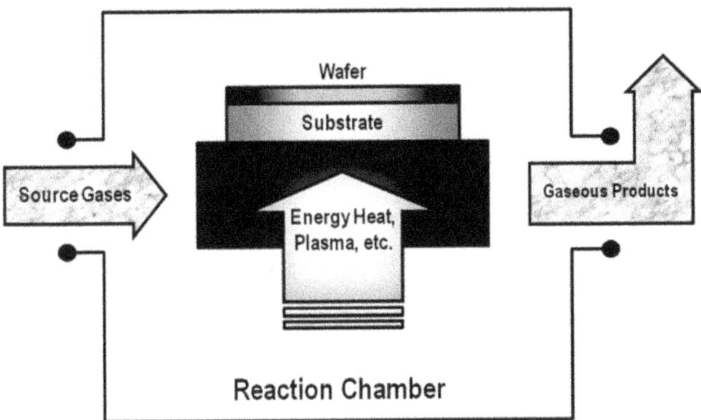

Figure 4.15. Diagram of chemical vapor deposition.

Zr/(Zr + Ti) variances of less than 5% using the MOCVD process. The AES depth profiling shows that the composition of the PZT films is consistent throughout their thickness.

4.8.2 Solution-deposition techniques

A solution-deposition technique enables excellent stoichiometric control of complex, mixed oxide systems and is a fast, cost-efficient method for preparing oxide thin films. The most commonly used solution techniques are sol–gel and metal–organic deposition (MOD). In solution deposition for PZT films, metal–organic starting reagents, such as alkoxide $M(OR)_n$ (where M is a metal element, and R is an alkyl group), carboxylate ($M(OOCR)_n$), and β-bi-ketonate ($MO_x(CH_3COCHCOCH_3)_n$) compounds of lead, zirconium, and titanium are widely used as precursor compounds.

4.8.2.1 Sol–gel technique

High chemical purity, strong stoichiometric control, low-temperature processing, simple thin-film deposition procedures, and microstructure make sol–gel technology one of the most common technologies for PZT thin deposition control via chemical reactions. The basic steps of sol–gel processing synthesize a gel-precursor solution, or sol–gel, which is applied to a substrate using a spin- or dip-coating technique. The pyrolyzed gel layer is then used to create an oxide ceramic coating. The use of sol–gel processes to make PZT films was initially reported by Fukushima and colleagues. They used butanol as the solvent for lead ethylhexanate, titanium tetrabutoxide, and zirconium acetalacetonate. Since then, great effort has been made to use the sol–gel technique to prepare PZT films. Thus far, sol–gel processes can be classified as (1) those based on the use of 2-methoxyethanol as a reactant and solvent, and (2) hybrid processes based on chelating ligands, such as acetic acid or ethanolamine, to reduce alkoxide reactivity. In 2-methoxyethnol processing, metal–oxygen–metal bonds are formed as follows:

$$M(OR)_n + H2O \rightarrow M(RO)_{n-1}(OH) + ROH$$
$$2M(RO)_{n-1}(OH) \rightarrow M_2O(OR)_{2n-2} + H_2O$$
$$M(OR)_n + M(OR)_{n-1}OH \rightarrow M_2O(OR)_{2n-2}(OH) + ROH$$

A key step in 2-methoxyethanol processing is the substitution of methoxyethoxy with Ti- and Zr- propoxide starting reagents. This substitution produces a less-sensitive precursor solution. The substitution is performed as follows:

$$M(OR)_n + nR'OH \rightarrow M(OR')_n + nROH$$

A hybrid method starts with alkoxide compounds but then adds other reagents like acetic acid and acteyleacetone diethanolamine to modify the reaction. Chelation of acetic acid is the most important reaction:

$$M(OR)_n + xCH_3COOH \rightarrow M(OR)_{n-x}(OOCCH_3)_x + xROH$$

Acetic acid makes the starting reagent less susceptible to hydrolysis and condensation. Figure 4.16 shows a flowchart depicting the various processing steps for manufacturing films (PZT). A typical sol–gel technique produces a single-layer PZT sheet with a thickness of 0.1 m. To obtain thicker sheets (0.5 m), multi-coating is commonly utilized. The updated sol–gel technique/process for PZT film deposition was by adopting a diol-based sol–gel procedure to achieve 0.5-m and thicker films in single coats. This modification to the sol–gel technique, enabling the production of thicker films, creates a significant advantage in film utilization in micro-actuators, pyroelectric sensors, and multi-layer capacitors.

4.8.2.2 Metal–organic deposition

For over a century, the MOD approach for thin-film production has been recognized. To obtain a genuine solution, the method begins by dissolving precursor chemicals in the appropriate solvent. After that, the precursor solution is spun coated or sprayed onto the substrate. The wet films are then heated to remove any remaining solvent from the deposition stage, and the organic films are pyrolyzed into inorganic films. If a thicker film is required, the coating and pyrolysis procedures are repeated until the necessary thickness is achieved. To produce the necessary

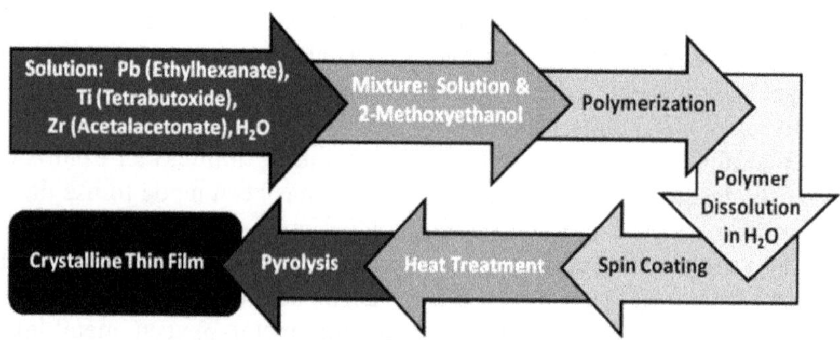

Figure 4.16. Flowchart of sol–gel processing of thin films.

crystalline phase, grain size, orientation, and oxygen content, the films are ultimately post-annealed. The following advantages of MOD are: (1) necessary cations are mixed at the molecular level; (2) low processing temperature; (3) a very stable precursor solution; (4) the deposition process may nearly always be carried out in air; and (5) facile application to substrates of any size and shape. For oxide film preparation, the MOD approach has recently emerged as a promising methodology.

Precursor synthesis

The ligands are bound to the central metal ion by a hetro-atomic bridge in the molecular structure of precursor compounds made with MOD. The bridge atom can be O, S, P, or N.

The following is a metal–organic compound diagram:

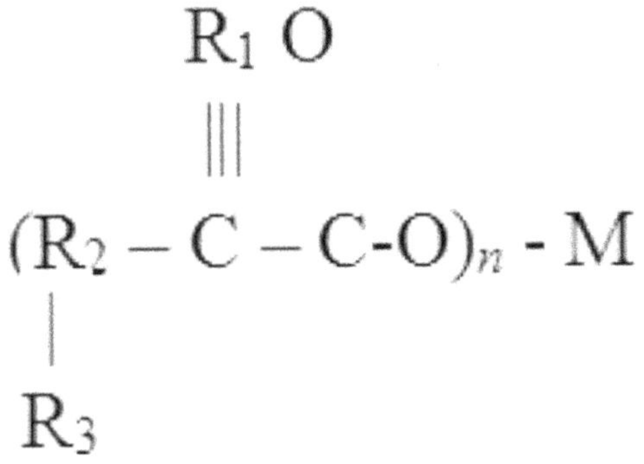

R1 O

|||

(R2 – C – C-O)$_n$ - M

|

R3

Because the majority of metal–organic compounds appropriate for oxide film production are not commercially accessible, the success of oxide film fabrication using MOD methods is highly dependent on either their availability or their synthesis. To date, a number of approaches for synthesizing various metal precursors to generate oxide films using the MOD process have been established. The primary methods are: (1) ammonium soap double decomposition; (2) neutralization; (3) metathesis reaction from a metal acetate; and (4) metathesis reaction from a metal alkoxide. In the MOD process producing oxide films, metal carboxylates are the most widely utilized starting ingredients. Metal carboxylate salts are made by combining a metal salt (usually hydrated nitrates or chlorides) with the creation of an ammonium soap of carboxylic acid, similar to double-decomposition processes. The procedure is as follows:

$$NH_4OH + RCOOH \rightarrow RCOONH_4 + H_2O$$

$$MX_n + n\text{RCOONH}_4 \rightarrow M(\text{RCOO})_n + n(\text{NH}_4)X$$

When the resulting metal carboxylate is solid particles, the powder is filtered and washed to separate the metal carboxylate from the homogeneous mixture. If the product is a liquid, a nonpolar solvent such as xylene or benzene can be used to extract the metal carboxylate from the aqueous solution.

Solvent

MOD involves dissolving metal precursors in a suitable solvent to create a liquid solution of organic molecules. The solvent must meet the following criteria: (1) high precursor solvency; (2) no precursor interaction; (3) adjustable viscosity; (4) acceptable vapor pressure; (5) low cost; and (6) non-toxic and non-corrosive properties. The solvents xylene and propionic acid are typically employed in the MOD process. Mixed solvents are used in situations where a single solvent does not meet all of the conditions. The overall solvency capabilities of a mixed solvent can be considerably improved by combining two or three solvents. the order in which the precursor solvents are added to the combination is critical throughout the precursor-solution formation process because of the micelle nature of the soap solutions. Micelles are thermodynamically stable collections of three or more soap molecules that exist in a liquid. The aggregation of polar carboxylate molecules can form a compact closed configuration instead of linear chains, resulting in a nonpolar structure and low viscosity carboxylate soaps in low polarity liquids. The capacity to dissolve some compounds by enclosing the molecules in the core of the micelles is an advantage of a soap solution's micelle nature in organic solvents. The molecules are not dissolved in the solvent if this is not the case. Lanthanum acetate, for example, is difficult to dissolve in xylene but dissolves quickly in a xylene solution of titanium di-methoxy-neodecanoate. The order in which the precursors are added is crucial since it determines what sorts of micelles are present in the solution and their solubility, as well as the precursors' solubility in the solvent. According to Wu and Sayer's PZT film research, zirconium propoxide must be added before titanium isopropoxide during solution preparation to avoid the creation of mono or diacetylated products, which would cause hydrolyzation to begin too quickly. Figure 4.17 depicts a typical technique for producing films via MOD.

Spin coating and pyrolysis

In the MOD process, spinning is the most popular approach for putting a homogeneous coating on a substrate. Because of the varied solubility of each metal–organic compound in the solvent, solvent evaporation of the precursor solution might induce inhomogeneity throughout the coating process, regardless of the solvent system. Thus, to avoid segregation, the deposition phase in the process should be completed as quickly as feasible. The spinning starts slowly to disperse the solution during spin coating. The spinning velocity is quickly increased to 3000 rpm for a longer duration

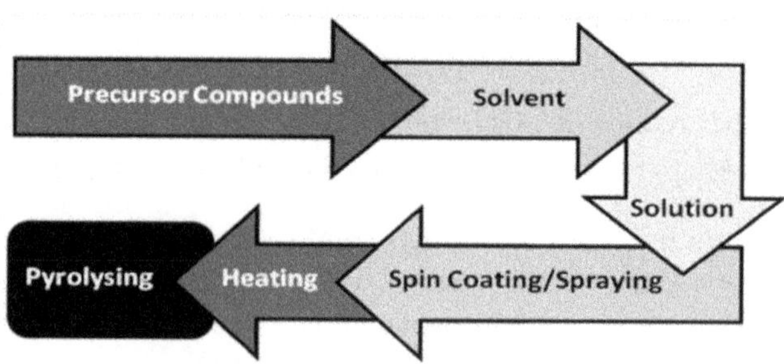

Figure 4.17. Flowchart for the MOD method.

once the precursor solution has completely covered the substrate (30 s to 5 min). Following spin coating, the green organic films are heat-treated at 300 °C–500 °C in air or oxygen to remove any remaining solvent not evaporated during the deposition stage and convert the organic films to an inorganic film. In a multi-component system, the homogeneous thermal breakdown is crucial to evade elemental segregation and generate uniform phases and characteristics. Because the microstructure of the film is produced during the pyrolysis process, it is the most crucial phase in the MOD process. The film's volume change is typically significant, with typical ratios of 6 to 30 for deposited thickness to burned thickness. During this step, an insufficient heating rate might cause faults in the burned film, such as cracks and voids. Regrettably, there is no guideline for determining an optimal heating rate. The thermal decomposition of carboxylates is thought to be a free-radical mechanism, with the decomposition reactions going through three stages: (1) the generation of free radicals via thermal fission; (2) the fragmentation and/or rearrangement of organic ligands; and (3) the oxidation of fission pieces of organic ligands via an oxidative chain reaction. The free radicals formed by the breakdown of the least stable metal–organic compound will cause the decomposition of the other precursors if the decomposition temperatures of the metal–organic compounds are not too variable. When compared to the temperatures of the individual components in the system, this impact will lower the total decomposition temperature.

4.9 Thick-film fabrication

4.9.1 Thick-film-transfer technology (screen printing)

Organic chemicals containing a solvent, dispersion agent, and plasticizer are first mixed with a binder and then blended in this type of screen-printing ink. After that, the active pyroelectric powder is added and stirred. The desired mask (screen) is placed overall in order to screen print, but the substrate on which the film is to be printed is not. A squeegee held at a 45-degree angle is placed on the screen behind the ink dollop and smoothly moved in an orthogonal direction across the entire substrate with minimal downward force. Following that, the screen is gently peeled away from the substrate. The sample is then taken from the fixture and allowed to rest for 10 min before being dried for 15 min on a hot plate set at 100 °C. To add further layers to the substrate, the process is repeated. Following the final coating, the film is sintered at a suitable temperature and for the necessary amount of time.

4.9.2 Design of thick film transfer apparatus

The apparatus for 'thick film transfer' (TFT) is shown in figure 4.18. The TFT system was fabricated according to the design and modified in the workshop. It consists of a base plate on which a vacuum chuck/substrate holder is placed at a predetermined location. The screen is placed close to the substrate, and ink/paste is placed on top of the screen. A soft rubber squeegee is moved across the screen and driving the paste in front of it and down through the unplugged areas of the screen onto the surface of a substrate. The screen is lifted off, leaving a patterned deposit of paste on the substrate, which is then dried, carbonized, and sintered to leave a thick film.

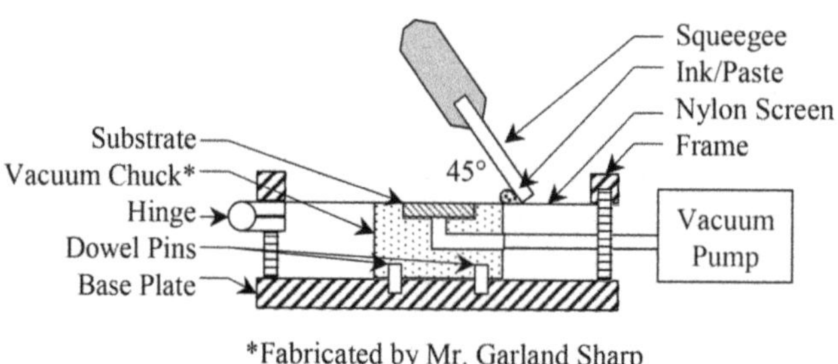

*Fabricated by Mr. Garland Sharp

Figure 4.18. A schematic of a screen-printing machine.

4.9.3 Thick film sample preparation methodology

First, a mask, a vacuum chuck, and substrates are fabricated. The mask is applied to a screen via a photolithography process using a template prepared by a computer and rendered on a laser jet (dry ink) printer. It is critical to ensure that the areas on the template are very black or very white because any voids in the printing on the template will surely transfer to the screen. A vacuum chuck is milled to hold and register each sample in the screen-printing fixture. A recess is milled in the top of the chuck, which is the same size and depth as the sample substrate size and thickness. A hole is drilled in the middle of the recess into approximately the center of the chuck. Another hole is drilled from the side of the chuck and connects to the hole drilled previously. A metal tube is pressed into the side hole and a vacuum hose is fitted. A 45° × 3 mm wide groove is milled on one side of the chuck to allow access for a tool to remove the sample from the recess. The vacuum chuck itself is also registered to the screen-printing fixture with two dowel pins. Two dowel pins were necessary to prevent rotation of the vacuum chuck on the screen-print fixture. The mask on the screen contains a series of circular openings with differing diameters and so the vacuum chuck is again registered to each position by drilling two corresponding holes for the dowel pins. The chuck is easily moved from one position to the next by simply pulling up the chuck and repositioning the dowel pins in the corresponding holes at the next location on the base of the screen-print fixture. Registration of each sample is necessary to allow more than one coat to be applied to the sample on the same location, thereby building up material on the surface of the substrate. A vacuum is necessary to hold down the sample while the screen is lifted off after the printing process. This eliminates the propensity of the sample to stick to the screen.

Second, substrates are made. Several types of sample substrates are utilized: stainless steel, nickel, alumina, and silicon. The substrates are cut from 1 mm stock material into 15 mm × 15 mm squares, which nicely fit into the milled recess of the vacuum chuck. The snug fit allows the sample to be registered on the screen such that additional layers can be applied with accuracy and precision.

The third step is to prepare the ink. The organic ingredients of the ink include solvent, dispersing agent, and plasticizer that were incorporated with the binder and mixed for 10 min. The active powder was added and mixed for another 5 min. The solvent terpinol is used to dissolve all constituents such that mixing is facilitated. The binder polyvinyl butyral adheres the mixture to the substrate. The dispersing agent buthoxyethoxy-ethyl adds to the rheology of the homogeneous mixture of 'ink'. The plasticizer polyethylene glycol reduces stress cracks when subjecting the film to heat treatments. The active ingredient PT or PZT has the necessary electrical properties that are under investigation.

Step four is to screen-print. The vacuum chuck is registered on the base of the screen print fixture to the desired unmasked feature of the screen with dowel pins and corresponding holes. A sample is placed in the vacuum chuck recess, a vacuum is applied, and the screen is lowered onto the substrate sample. A dollop of ink is placed on the screen at a point behind and stretching in an orthogonal direction across the unmasked feature to be printed. A squeegee is placed behind the ink dollop, onto the screen, held at a 45° angle, and then in a smooth motion with minimal downward force; the squeegee is gently drawn towards the body and across the top of the unmasked screen feature. The screen is gently lifted off of the sample substrate, and the vacuum is released, the sample is removed from the fixture, and allowed to rest for 10 min. Next, the sample is dried on a hot plate set to 100 °C for 15 min. The process is repeated to add additional layers onto the substrate. The complete process is illustrated in the flow diagram shown in figure 4.19. A Typical composition of paste used in the preparation of various films is tabulated in table 4.3.

Step 5 is to administer two heat treatments. The first heat treatment is called calcification, whereby the sample is baked in a furnace at 500 °C for 5 h then allowed to cool down slowly to ambient conditions. The second heat treatment called sintering requires that the sample be baked in a furnace at elevated temperatures for 5 min and again allowed to cool back down to ambient conditions. Several sintering temperatures were chosen: 700 °C, 800 °C, and 1000 °C.

Step 6 is to apply an electrode to the surface of the sample. A sample is placed in a die of known diameter, and a spot of silver paint is applied to the surface of the thick film and allowed to dry for ~ 24 h. A continuity test is employed to confirm that the silver paint is not shorted out to the surface of the substrate and that the sample is viable for testing by virtue of the presence of high resistance from the silver paint through the film to the substrate surface.

4.10 Fabrication of polymer–ceramic composites

The manufacture of 0–3 connectivity composites is straightforward, enabling for cost-effective commercial production of these materials. As a result, the vast majority of composites are made as active pyroelectric and polymer-based diphasic samples. This section describes composites that have simple composite-fabrication approaches that have provided some promising results. Polymer components can be either polar or nonpolar. The sol–gel synthesis method is a second prominent method. Although notable pyroelectric sensing elements have been shown in glass-ceramics, this section does not include precise descriptions of them. In a hot rolling mill, ceramic particles are combined with a softened thermoplastic polymer, and thin composite films are produced by high-pressure casting at the polymer's softening temperature. It is feasible to mix a thermoset polymer like epoxy with the right quantities of resin, hardener, and ceramic powder at room temperature. In solvent-casting, a polymer is dissolved in a

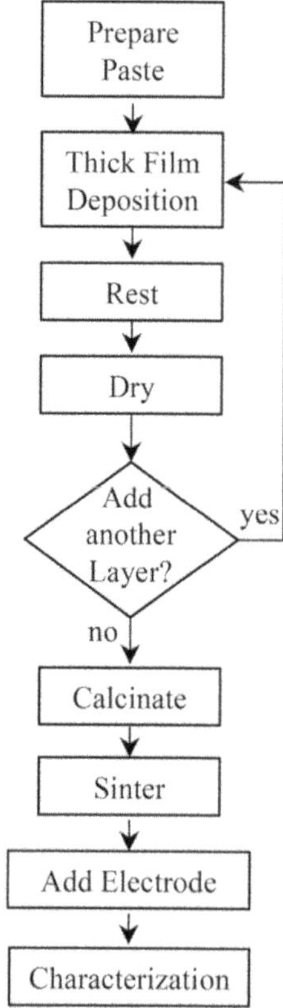

Figure 4.19. The complete process is illustrated in the flow diagram.

Table 4.3. A typical composition of paste.

Material	Purpose
Terpinol	Solvent
Polyvinyl butyral	Binder
Buthoxyethoxy – ethyl acetate	Dispersing agent
Polyethylene glycol	Plasticizer
PT or PZT	Active ingredient

suitable solvent before an electro-active ceramic powder is added and mixed/dispersed. The mixture is kept in a container with a vent that allows the solvent to escape. The film is then heated to the crystallization temperature of the polymer. During the mixing process, issues such as poor ceramic inclusion distribution, component phase adhesion, and air bubbles in the composites are addressed. The problem of ceramic particles agglomerating in the solution can be solved by reducing the viscosity of the polymer during the ultrasonication process. The following is a typical technique for preparing a P(VDF-TrFE):PZT composite and using spin coating to fabricate 0–3

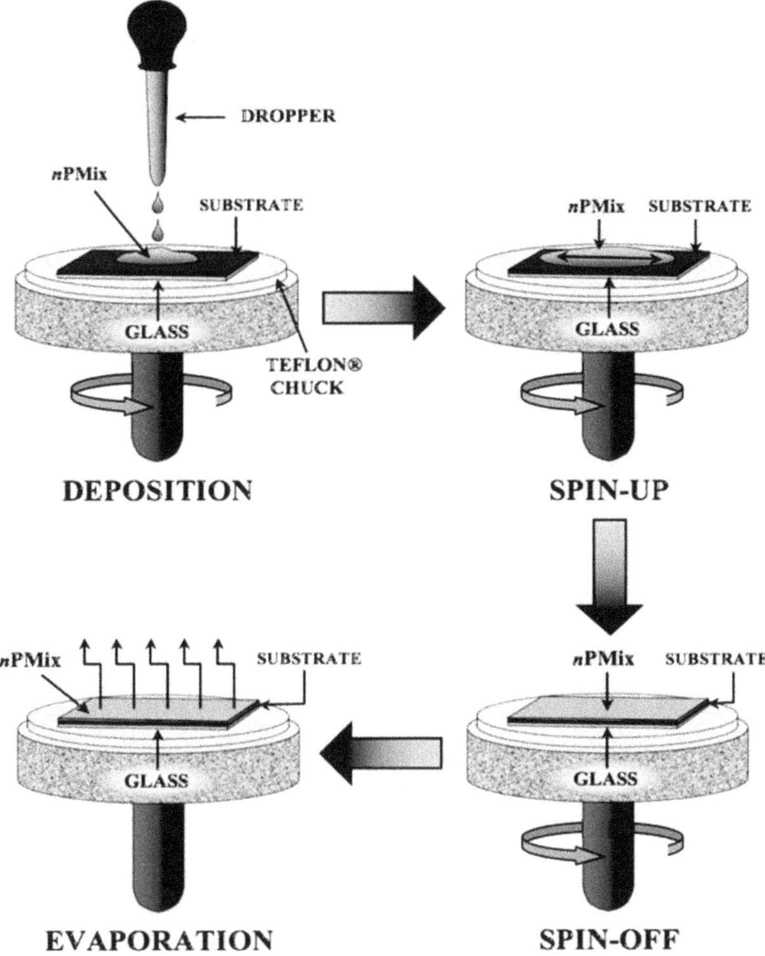

Figure 4.20. Spin coating technique processing steps.

connectivity composites: A suitable amount of the polymer P(VDF-TrFE) is dissolved in methyl-ethyl-ketone (MEK) to form a solution (PMix),

1. A requisite amount of nano-ceramic (PZT) powder is added.
2. The mixture is ultrasonically agitated for several hours to break up the agglomerates and disperse the ceramic powder uniformly in the copolymer solution;
3. Using this composite solution (nPMix), a thin film is deposited on a conducting electrode substrate using a spin-coating technique;
4. The film is annealed for 2–3 h in the air at 130 °C; and
5. The top electrode is deposited on the film for testing.

Figure 4.20 depicts the methods involved in producing films with nPMix by spin coating. Figure 4.21 depicts a flowchart depicting the various procedures required in producing composite films.

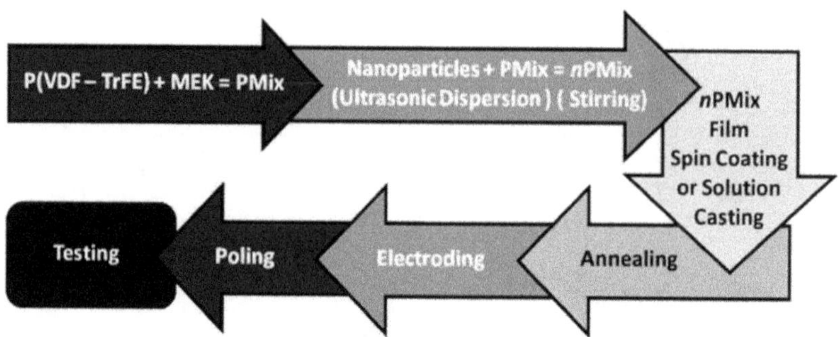

Figure 4.21. Flowchart for fabricating polymer–ceramic composite films.

References

[1] Batra A K, Mohan D and Aggarwal M D 2018 *Field Guide to Crystal Growth* (Seattle, WA: SPIE Press)

[2] Lecoq P, Gektin A and Korzhik M 2006 *Inorganic Scintillators for Detectors Systems* (Berlin: Springer Publishing)

[3] Gupta T K 2013 *Radiation, Ionization and Detection in Nuclear Medicine* (Berlin: Springer Publishing)

[4] Owens A 2016 *Compound Semiconductor Radiation Detectors* (Boca Raton, FL: CRC Press)

[5] Awadalla S 2015 *Solid-State Radiation Detectors* (Boca Raton, FL: CRC Press)

[6] Dujardin C, Auffray A, Bourret-Courchesne E, Dorenbos P, Lecoq P, Nikl M, Vasil'ev A N, Yoshikawa A and Zhu R Y 2018 Needs, trends, and advances in inorganic scintillators *IEEE Trans. Nuc. Sci.* **65** 1977–97

[7] Greskovich C and Duclos S 1997 Ceramic scintillators *Annu. Rev. Mater. Sci.* **27** 69

[8] Koshimizu M 2016 Sol–gel-doped glasses for scintillators *Handbook of Sol–Gel Science and Technology* (Berlin: Springer International Publishing)

[9] Arshak K and Korostynska O 2005 Thin- and thick-film real-time gamma radiation detectors *IEEE Sens. J.* **5** 574

[10] Milbrath B D, Peurrung A J, Bless M and Weber W J 2008 Radiation detectors materials: an overview *J. Mater. Res.* **3** 2562

IOP Publishing

Advanced Nuclear Radiation Detectors

Materials, processing, properties and applications

Ashok K Batra

Chapter 5

Application of gamma radiation detectors and outlook

5.1 Introduction

Scintillating materials are now widely employed in a variety of detecting systems in a variety of fields. Medical imaging, homeland security, nuclear nonproliferation, HEP calorimetry, industrial monitoring and control, oil drilling exploration, astrophysical study, and basic research are among the fields. Inorganic materials account for a significant portion of research and market share, with a market value of $350 million in 2015. By 2022, the global radiation safety market is expected to be worth $2.26 billion. Mounting security threats, rising cancer prevalence worldwide, increasing safety awareness among people working in radiation-prone environments, growing safety concerns, growing security budgets of global sporting events, growth in the number of PET/CT scans, and increasing use of nuclear medicine and radiation therapy for diagnosis and treatment are the key factors driving the growth of this market.

More than a century after the initial application of scintillating material, research is still highly active as detector technologies advance and ionizing radiation systems' capabilities and performances change. Furthermore, material synthesis processes and related knowledge have improved and expanded significantly. Since then, there has been tremendous progress in the theory and modeling of scintillation mechanisms. The latest advances and trends in inorganic scintillation science are discussed in this chapter. Emerging fields of interest in scintillation as it relates to materials are of particular interest. The new materials will not be able to replace commonly used compounds like NaI: Tl, CsI: Tl, and Lu_2SiO_5:Ce. However, when compared to well-established compounds [1], it is believed that the reported locations would be able to outperform some of the quality standards.

doi:10.1088/978-0-7503-2508-0ch5

5.2 Desired properties of scintillators and respective applications

Some qualities are optimized by adjusting the application of scintillators. A good scintillator has strong light output, exceptional energy resolution, a fast scintillation decay time, resistance to radiation damage, high density, and emission in the photodetector spectral response. Table 5.1 shows the relationship between the scintillator parameters and the application.

5.3 Applications of scintillator crystals

The last decade has seen a renaissance in inorganic scintillator development to detect various kinds of ionic radiation. Inorganic crystalline scintillators are applied in devices for high energy physics, tomography, astrophysics, solar neutrino detection, well oil-logging, etc.

5.3.1 Medical applications

Medical imaging is a rapidly evolving profession. It is based on five different modalities: x-ray radiology, emission tomography, ultrasonic tomography, magnetic resonance imaging (MRI), and electrophysiology (EEG and MEG) [2].

The stopping power for the given energy ranges of x- and γ-rays to be considered, and, more accurately, the conversion efficiency, are the first crucial requirements for a scintillator to be employed in medical imaging equipment in concept. Materials with a high Z and density are clearly preferred, but the position of the K-edge is also critical. The attenuation coefficient of yttrium, cesium, and iodine is rather high for low-energy x-ray imaging (below 63 keV), therefore crystals like YAP and CsI are

Table 5.1. A list of prefered properties, relevant applications, and possible materials.

Prefered properties	Applications	Possible major material
Very high light, good energy resolution, large crystal size	Tomography. Counting, medical physics, security, space, environmental monitoring	NaI(Tl)
High light output, rugged	Tomography, geophysical, radiation detectors	CsI(Na)
Fast, non-hygroscopic	Calorimetry, high energy physics	CsI
Very high light output, very good resolution	Spectroscopy, medical physics, environmental monitoring, space	$LaCl_3$:$Ce_{(0.9)}$
High density and Z, fast	Space, physics research, PET, high energy physics	LYSO
High density and Z	Nuclear physics, geophysical research, PET. Security (isotope identification)	BGO
High density, fast, low afterglow	High energy physics, calorimetry	$PbWO_4$
Fast, low density and Z, light output	Tomography, neutron detection, general counting	Plastics

suitable choices to utilize for each application. However, because the ideal scintillator does not exist, each modality must make a compromise, selecting the best combination of features from the available materials. A wide range of medical imaging equipment uses scintillating materials. Inorganic scintillators have traditionally played a key role in the detection and visualization of radiation.

5.3.2 Homeland security safety system applications

Scintillators are used in three different types of detecting equipment. These include express luggage and passenger control, explosives check, and fissile material identification from afar. The goal is to quickly identify a suspected piece of luggage in a container that is a few cubic meters in size and traveling over the inspection equipment. The highest possible throughput is a crucial need for such a scanner. The necessity to locate and maybe identify the suspect article in a large container is related to the spatial resolution. Nuclear radiation detection methods appear to be very promising for the remote detection of explosives. These approaches rely on the detection of distinctive neutron or γ-rays, which can be natural or produced.

5.3.3 Astrophysics

Scintillators are used in space physics in two different places: low-orbit satellites and space or interplanetary missions. The Earth's magnetic field shields low-orbit satellites, allowing the requirement for radiation hardness of the scintillation material to be relaxed. Depending on the energy range of the observed γ-radiation, most scintillation materials can be employed. The payload, on the other hand, limits the size of such detectors, therefore low-density materials are sometimes chosen to save weight. The detecting properties of scintillation materials are substantially influenced by the Sun's charged particle stream in interplanetary space [3]. High radiation hardness to ionizing radiation and a low amount of generated radioactivity is required for these missions. The same can be said for the detectors that are sent to the planets. Although research in this area is still in its infancy, it is reasonable to conclude that scintillators that are reasonably light, fast, and bright are the most promising for future space missions and that LaBr3, YAP, and (Lu–Y) AP are likely to become the scintillators of choice [4].

In the recent past, a portable Si/CdTe Compton camera and its application to the visualization of radioactive substances have been developed [5]. It can be used to visualize the distribution of radioscopes in many fields such as scientific research, non-destructive inspection, and homeland security.

5.3.4 Oil and natural gas exploration and production technologies

The goal is to create a novel solid-state gamma-ray detector that can be used in difficult environments throughout the gas and oil exploration process. The major outcomes were achieved by a combination of tasks. Avalanche photodiode (APD), scintillator, optical coupler, and electronics are all key detector components that have been successfully created. Fabrication and testing of prototype systems has taken place. The results showed that the HPHT sensor system has insufficient system performance to be commercialized at this time. The initial speculation is that the APD electrical leakage was too high, and that the scintillator development of a high-temperature, solid-state photodetector fabricated from silicon carbide will overcome current gamma detector temperature limitations, as silicon carbide devices can survive and perform at temperatures above 200 °C. In comparison to present detectors, the new detector will have two major advantages. For starters, it will be able to work at higher temperatures, allowing for deeper drilling and exploration. Second, the solid-state photodetector will have a longer life down holes because it is more resistant to shock and vibration. Both of these benefits will aid in lowering the risk and cost of deep drilling.

5.4 Outlook and future prospects

For inorganic scintillator materials, single crystals are still the most common bulk form. However, the limitations imposed by the manufacturing process have enticed researchers to seek out and develop alternative media, such as transparent ceramics, particularly for applications requiring large volumes or areas, and composite materials in an attempt to couple specific functions of the various phases. We will examine recent advancements and trends in inorganic scintillators grown as single crystals and created as transparent ceramics in this part. The Czochralski process, which is used for large-scale production, is used to create oxide crystals in general. The method is entirely automated, the crystal growth is monitored in real time, and the production yield is great. Iridium crucibles are widely employed for high melting point oxides, which require very little oxygen in the growing atmosphere.

Many studies targeted at improving the yttrium aluminum garnet YAG:Ce have yielded an improved garnet by substituting Gd for Y and alloying Al with Ga. The inclusion of gallium oxide hampers the development process because it requires a low quantity of oxygen to avoid breakdown. However, solitary crystals of Ce-doped garnet $Gd_3Al_2Ga_3O_{12}$ with a diameter of 3 inches have been recorded [6]. These Ce-doped garnets are now commercially accessible, and the use of co-dopants allows for the customization of one or more parameters, such as timing, light output, and energy resolution.

In the case of halides, scale-up attempts for the binary compound SrI_2:Eu, which is now commercially accessible, have proceeded. Because the material is liquescent, adequate handling and raw material quality have been investigated [7, 8]. With two significant exceptions, the Bridgman approach is used in the majority of endeavors.

The Czochralski process has been used to demonstrate the growth of that material at diameters up to 50 mm.

A new growth approach, similar to a seeded vertical Bridgman technique, employs a graphite crucible in an evacuated dry chamber, allowing for high yield production.

Despite the fact that NaI:Tl and $LaBr_3$:Ce are grown in production plants using the Czochralski technique, the newly discovered multi-component halide scintillators are primarily grown using modified Bridgman–Stockbarger techniques due to their reactivity and hygroscopicity, which necessitates the use of chemically compatible crucibles and a dry atmosphere. Because of its simplicity, the Bridgman–Stockbarger technique is simple to adopt in a research laboratory with modest initial expenses. However, the approach has a low yield in general, particularly when employed without a seed, which is how the majority of research is done. Contamination, crystal clinging to the ampoule, and splitting can be caused by a confinement in a crucible (typically quartz that can be easily sealed) or vitreous carbon put in a quartz ampoule. The lack of reproducibility and low yield has been extensively established, as evidenced by numerous articles, such as attempts to generate multiple crystals in one furnace [9]. As a result, scale-up efforts for these newly found scintillators are currently limited. The majority of novel multi-component materials are generated in sizes smaller than 25 mm in diameter. We must note that a few attempts at growing these compounds by the Czochralski technique have demonstrated that it is viable and should be pursued [10].

5.4.1 Transparent ceramics

When a precise geometrical form is difficult to accomplish with single crystals and they are made at a lesser cost, transparent ceramics are frequently used as a substitute. Transparency, on the other hand, denotes a lack of light scattering, which can be caused by minor flaws such as micropores or grain boundary disruptions. When birefringence develops in noncubic materials, it can produce scattering and must be regulated by grain size. Vherepy *et al* [11] showed a successful example of improved transparent GYGAG ceramic. The cubic material LuAG:Ce showed structural disruptions at grain boundaries, indicating that transmission of transparent oxide ceramics can still be enhanced. Because of their reactivity, fabricating translucent ceramics using hygroscopic halide materials is more difficult, and few results have been documented. SrI_2:Eu has been attempted as a translucent ceramic with minimal success [12]. SrI_2 is orthorhombic, and scattering difficulties have yet to be resolved [13].

5.4.2 Codoping

The atomistic perfection of the scintillator material becomes crucial because the scintillation mechanism comprises a transfer stage in which migrating electrons and holes in the conduction and valence bands, respectively, must overcome impediments on their journey to reach the emission center. Even in high-quality single-crystal hosts, point defects, primarily cationic and anionic vacancies, are unavoidable, resulting in hole and electron traps, respectively. Charge traps can also be caused by other types of lattice disorders, such as unintentional impurities and more extensive lattice faults (such as dislocations). Because manufacturing techniques alone cannot completely eliminate these flaws, new methods have been created and are being developed to improve scintillator qualities as necessitated by specific applications. Co-doping of the scintillator material, which involves the addition of a specific impurity that may (or cannot) participate in charge carrier capture, is one of these tools. The co-doping of Gd_2O_2S:Pr, phosphor, or ceramic by Ce^{3+} and F^- ions which successfully reduces the afterglow and allows it to be used in computed tomography (CT) medical imaging [14], is a well-known example.

5.4.3 High-*Z* sensitized plastic scintillators

Plastic scintillators are a cost-effective way to build a large-volume detector. However, the small radiation stopping power inherent in their low atomic number significantly limits their performance for high-energy gamma radiation. With the advancement of organic optoelectronics and nanotechnology, interest in heavy element sensitized plastic scintillators has recently resurfaced. Readers can consult the references for more information on physics and other topics. It is proposed that a nanocomposite based on doped luminous inorganic nanoparticles be produced and that this could be a promising topic for further research and development [15].

5.4.4 Perovskite single crystals

Many efforts have been made to minimize the cost of detectors due to compositional non-uniformity and large concentrations of Te inclusions in CdZnTe crystals, as well as the high fabrication cost of detectors. The quest for novel semiconductor gamma-ray detector materials to cut production costs while maintaining equivalent device performance is constant—the perovskite family is one such a group of contenders with considerable potential. It has a general chemical shape ACX_3 (where A = Cs+, methylammonium (MA or CH_3NH^{3+}), formamidinium (FA or $HC(NH_2)^{2+}$), Rb (MAFA), B = Pb^{2+}, Sn^{2+}, X = Cl^-, Br^- I^-) that has gotten a lot of attention in recent years for applications in optoelectronic devices, light-emitting diodes, and radiation. Single crystal-based Schottky detectors based on $MAPbI_3$ and $CsPbBr_3$ have been successfully confirmed for gamma spectroscopic usage. Even though Perovskites' current energy resolution is lower than that of CdZnTe and CdZnTeSe detectors, their low fabrication cost, high average atomic number, and low density of trap states make them promising next-generation semiconductor candidates for gamma photon detection. However, there are a few obstacles to overcome in order to encourage the development of perovskites, such as to address the instability issue of the Pb component, Cs A-site cation alloying could be used; additionally, to avoid toxic Pb, all inorganic double perovskites (e.g. $Cs_2AgBiBr_6$, Cs_2PdBr_6, $MATlBiBr_6$) could be used; and structural engineering could be used to suppress the undesirable ionic migration.

References

[1] http://www.hep.caltech.edu/~zhu/papers/18_tns_scint_all.pdf

[2] http://www.aessweb.com/pdf-files/JASR-131-136.pdf

[3] http://lib.convdocs.org/docs/index-125468.html

[4] https://epdf.pub/inorganic-scintillators-for-detector-systems-physical-principles-and-crystal-eng.html

[5] Takeda S and Harayama A *et al* 2015 A portable Si/CdTe Compton camera and its applications to the visualization of radioactive substances *Nucl. Instrt. Meth. Phys. Res. A.* **787** 207–11

[6] Kamada K *et al* 2016 Growth and scintillating properties of 3 in. diameter Ce doped $Gd_3Ga_3Al2O_{12}$ scintillating single crystal *J. Cryst. Growth* **452** 81–4
[7] Gektin E *et al* 2016 Strontium iodide:technology aspects of raw material choice and crystal growth *Funct. Mater* **23** 473–7
[8] Hawrami R *et al* 2008 SrI_2: a novel scintillator crystal for nuclear isotope identifiers *Hard X-Ray, Gamma-Ray, and Neutron Detector Physics* Xvol 10762 (Bellingham, WA: SPIE)
[9] Lindsey A C *et al* 2017 Multi-ampoule Bridgman growth of halide scintillator crystals using self-seeding method *J. Cryst. Growth* **470** 20–6
[10] Yan Z 2016 Shalapska and Bourret Czochralski growth of mixed halides BaBrCl and BaBrCl:Eu *J. Cryst. Growth* **435** 42–5
[11] Vherepy N J *et al* 2014 High energy resolution with transparent ceramic garnet scintillators *Proc. SPIE* **8213** 921302
[12] Podowitz S R *et al* 2010 Fabrication and properties of translucent SrI_2 and Eu:SrI_2 scintillator ceramics *IEEE Trans. Nucl Sci.* **57** 3827–35
[13] Dujardin C *et al* 2018 Needs, trends and advances in inorganic *IEEE Trans Nucl. Sci.* **65** 1977–97
[14] https://advances.sciencemag.org/content/advances/6/35/eaay4922.full.pdf
[15] Jacob T, Chao L, Cherepy N J and Pei Q 2018 High-*Z* sensitized plastic scintillators: a review *Adv. Mater.* **1706956** 1–13
[16] Zheng Z and Yang G 2020 Recent advancements in using perovskite single crystals for gamma-ray detection *J. Mater. Sci.: Mater. Electron.* **32** 12758–70

Lightning Source UK Ltd.
Milton Keynes UK
UKHW031441221021
392663UK00004B/98